A FEW GOOD OLD MEN

Books by Kee Briggs

The Third Removed
The Painted War
Finders-Keepers
Losers-Weepers

The Usher Orlop Mysteries

The Golden Janus
The Pewter Masks
The Nickel Trophy
The Bronze Bones
The Brass Portraits
The Zinc Ormolu
The Silver Scepter
The Rhodium Dragon

The Asti Fantasies

Charm Catcher
Dream Weaver

Ebook
Writer to Live Longer

A Few Good Old Men

Kee Briggs

Keescapes Publishing

Satellite Beach, Florida

A Few Good Old Men

Keescapes Publishing books may be ordered through booksellers or by contacting:

Keescapes Publishing

90 Flamingo Dr.

Satellite Beach, FL 32937

www.keescapes.com

keescapespublishing@gmail.com

This is a work of fiction. All the characters, names, incidents, organizations and dialogue are all figments of the author's imagination or used fictiously.

ISBN 978-0-9820044-3-2

Published in the United States of America

A Few Good Old Men

Kee Briggs

CHAPTER 1

Chucky maneuvered his wheelchair so that he could reach the key card slot. Getting into the American Legion hall was always a chore. Usually, one of the guys would see him and come to help manage the two pairs of spring-loaded double doors. This time he had to thump and bump his way in.

The reason no one had noticed him was that all eyes were focused on the TV at the end of a horseshoe bar. Whatever had been attracting everyone's attention had finished by the time Chucky made it in. The news cycled on.

Chucky rolled to his usual space at a large round table near the back of the dining area. There was no chair in his slot at the back, facing the door. Babs, the buxom blonde barmaid almost beat him to the table with his Coors.

"Hi love, don't give me any guff. Vengeance could be mine."

Babs slid the beer across to him and with a straight, stiff back marched back to the bar.

Chucky's eyes followed the receding swishing butt. "What was that all about?" he thought.

Arn, an old Korean vet, detached himself from a clutch of

drinkers and headed across the dining room. "Mornin', Chucky. How're you doing today?"

"Pretty good, Arn. What's with Babs and this vengeance bit?"

Arn cackled. "You didn't see that TV thing?"

"No."

"There was a news story about a woman taking vengeance on her abusive husband. He made a habit of coming home drunk on Saturday nights. He'd beat on his wife a while and then go flop down on the bed and pass out.

"Last time he pulled that stunt, he didn't notice the new bed-spread. It was made of heavy canvas with grommets around the edge. When the joker passed out, his wife tossed the 'bedspread' over him and pulled a drawstring, sealing him into a pocket. She then reached under the bed for a 2x4 and proceeded to beat the piss out of him. She dropped the 2x4, collected her suitcase from the hall closet, picked up her husband's car keys and left town.

"It wasn't until the joker didn't show up at work on Monday that a friend got worried. When there were no answers to the phone calls, he dropped by the house and found the wife beater. His wife had really wumpted him a goodie. There were some bro-ken bones and everything else was black and blue. He didn't smell pretty either."

"What's going to happen to the wife?"

"Probably, nothing. The husband was a notorious wife beater. The SA didn't want to pursue the matter in an election year and lose all of the female vote."

After all that talk, Arn needed to wet his parched tongue. He had to use both of his gnarled, arthritic hands to pick up the beer can that Babs had just delivered. After a prolonged draught, Arn sighed. "Here it's only goin' on noon and that first big swig will be the most exciting event of the whole darn day. From here on out, everything will be a downhill slide."

"It can't be that bad," said Chucky.

"You are young. You don't know what it's like getting old and infirm." Arn caught himself and tried to talk his way out of it. "Just like I don't know what it's like not to have legs." said Arn lamely.

Chucky took a sip of beer to avoid pursuing that line of thought.

Another of the old regulars came into the club, motioning to Babs before taking a seat at the table.

"How's it today, Hep?" said Chucky.

"Just variations on the same old theme. Had to have another blood test, which means I had to skip breakfast. He raised the Bud that Babs had brought. "The first sustenance of the day."

The next arrival was Merv, one of two the younger ones and by far, the heaviest of the Pessimists' Roundtable, as the other legionnaires called the group. It took a while for Merv to settle in. He pulled two chairs away from the table and replaced them with a broad chair big enough to be considered a loveseat. Conversation was put on hold until the shuffling and bumping and groaning settled down. Merv finally plopped into his place across from Chucky. Talk resumed over his wheezing and sucking for air.

Every day Merv felt put upon because no one ever offered him any help getting settled. Some respect should be shown for anyone with his disabilities. It seemed as if it was only at the Legion that he had to fend for himself.

Of course, it was the guys at the Legion who got a perverse pleasure watching that hulk flounder around. It was their opinion that his disability was a self-inflicted wound. Merv was a local boy, so the story was well known. He'd enlisted and gone to Nam, for which he was given credit but after that, everything went south. He got hooked on drugs and brought the habit home. Then he spent the next phase of his life eating and drinking himself into a blimp, accompanied by all of the associated ailments, such as diabetes.

All of his doctors, family and friends tried to get him straightened out. His only concession to his blimp status was to switch to light beer. He'd admit to 550 pounds, which was either a lie or he hadn't found a scale big enough to weigh his carcass in the last couple of years.

"What's the blood test for, Hep?" said Chucky. "You just had one the other day."

"Yeah, those quacks have got me on so many drugs they're fighting each other. They make me feel worse than the problem they're trying to correct. I've been bitching about the side effects and the cost. My health plan doesn't recognize a couple of the new drugs so I've been paying full price. I can't afford that. The doc is going to reevaluate my drug list"

"I know what you mean," said Arn. "If I didn't have the VA, I'd never make it."

"At least you haven't had to switch to draft beer," said Will, the latest addition to the roundtable. "Those drugs can kill you. Are there any afternoon games?"

"No. I checked," said Chucky. "White Sox play tomorrow at 2:00."

"That's better than nothin'. Where's Dane? He usually beats me here," said Will.

"He's having a bad breathing day," said Chucky. "He may not make it today. He's not doing too well."

Arn was facing the front window. "Here he comes now. His car just pulled into the lot."

Babs also noted Dane's arrival. She had a vodka on the rocks at his place before he could make it in the door. He always stoked up on oxygen before entering. He was afraid to bring his oxygen bottle in with so many smokers waving cigarettes around.

Dane saved his breath and just waved at the guys before dropping into the last chair at the table. It took him a while to catch his breath. After taking a sip of vodka, Dane was ready to become an active participant in the proceedings. "Did you see that news report about the wife getting retribution on her abusive husband?"

"It was on the tube just before I got here," said Chucky. "Arn told me about it. Tell the rest of the guys."

Dane told his story with a few personal embellishments.

"Good for her," said Hep. "I think this world would be a lot better place if there was more attention paid to paybacks. Maybe some of those yahoos would think twice before pulling a shitty on people if they knew it was going to come back to bite them."

"Retribution ought to be in spades," said Merv. "The payback should be a multiple of the original sin."

"That's what religion is supposed to do," said Dane.

Merv shrugged. "The problem with that is, you never know if God ever got the job done. I'd rather there was a more local enforcement."

"How can you figure the damages?" said Hep. "You'd have to know the sum of the losses and then you have to factor in some sort of retribution quotient. I'm afraid it would all get pretty complicated. You'd need to be part accountant and part God."

"I could handle that," said Will, "if I could get that son of a bitch that killed my grandson."

"How's that?" said Chucky.

"It happened just before you got here," said Hep, the old-timer. "We never bring it up. It's too fresh."

"Sorry."

Will batted back the tears. "That's all right. I should be getting over it by now. Don't feel bad. I brought it up because it's a prime example of where retribution is needed. Law won't do anything and God hasn't evened the score so that black bastard is still walking around free and probably endangering other kids.

"Danny was a good kid. Like all teenagers, he was going through his defiant period. You know...his parents were stupid and just didn't understand. He could take care of himself...make his own decisions, run with whomever he pleased, come and go as he wanted. You know the routine.

"My son and his wife were trying to work through the problem. Danny started running with a new, wild crowd. Willy, my son, grounded him, but then he sneaked out of the house to go to a big party thrown by this black kid, Mo.

This Mo is the local kids' drug dealer. In this case, retribution should carry on to the parents. They both work for the government. They were usually assigned overseas. When their contracts terminated, they decided to stay stateside and get their kid into American school. Apparently, that didn't work. The kid was so out of control, the schools wouldn't put up with him. The parents were afraid of him too. He had more money than they did from selling drugs. He started carrying two guns. At least that's the story that got pieced together and was passed on to me.

"Anyway, Danny showed up at the party and started using some bad drugs. Several of the kids got sick but everyone was afraid to call for help. Mo drove the sick ones out of his house. Danny passed out. Somehow the other kids dumped Danny on the lawn at his own house. Willy found his son's body when he went out to get the morning paper." Will stopped to regain some composure.

"Did the police scoop that Mo up?" said Chucky.

"Hell, the police asked a few questions and then dropped the case. They said they couldn't waste their time on a case that would never be prosecuted. Mo is only 15...a juvenile. They'd dealt with him several times already and no one could do anything. They were content to wait until they could prosecute him as an adult."

Chucky looked incredulous. "What about the parents? They have some liability."

"As the story goes, Mo threatened them. They took off overseas again and left him in the house. The FHA will eventually foreclose on the house, but that will take an eternity, especially when local representatives have been threatened personally if they do anything."

"So he's getting away with Danny's death?" said Chucky.

"Yeah, and that kid is still pulling all sorts of crap. Everyone just looks the other way."

There was a period of silence around the table. Merv motioned for another beer. Babs polled the group and filled the orders.

"Yeah, that would be rough," said Chucky. "This is a case where no justice has been served."

"That's what's wrong with the world," growled Hep. His "first sustenance" was turning him into a philosopher. "People are taking more and more liberties and not getting jerked up for it. That's just like the drunken doctor who killed my wife. Nothin' ever happened to him. If one of us would be picked up for drunken driving, we'd pay a lot more than that son of a bitch. Before I die I should go get him."

"Yeah, but if you do, you can bet that officialdom would come after you," said Merv. "It's like in football, the ref misses the first personal foul, but always slaps the foul on the guy who retaliates."

Hep shrugged his emaciated shoulders. "It only means anything if you've got something to lose. What have I to lose?"

There was an edgy silence around the table. Most attended to their drinks. Chucky shifted his weight to ease some pain.

Hep didn't get an answer, so he repeated his question, "What have I got to lose?"

"I guess none of us has much to lose," said Chucky. "At least for some, there may be family considerations."

"I've already outlived anyone who would care," said Hep. "If none of you guys come to my funeral then there wouldn't be anyone there."

"I thought you said you were going to be cremated," wheezed Dane.

"I am, but you know what I mean. Years ago I went to the funeral of my brother-in-law. There were five of us at the service, me, my wife, my father-in-law, and Rich's aunt and uncle. Across town, they were burying Walt Disney. The whole world mourned.

I am the last one left from that funeral. When I die, there will be no one left that will remember that Rich ever passed this way."

"My parents are still alive," said Merv, "but I wouldn't invite them to my funeral. They probably wouldn't come even if they were invited. We still don't see eye to eye. They never forgave me for getting hooked on drugs and then when I put on all this weight, they blamed it all on my weak character. They wouldn't give me even an ounce of credit that this weight gain was started by that crooked doctor."

"How's that?" said Chucky.

"The doctor I was seeing was writing prescriptions for me on a drugstore on the ground floor of the building. It comes out that the doctor had a major financial interest in the store. The druggist was buying black-market, outdated drugs. Some of those had lost their potency and others changed their characteristics. I became fat and diabetic and I never had the intestinal fortitude to correct the problem when there was a chance of doing it."

"Did they ever do anything to the doctor?" said Arn.

"No, not really. They closed the drugstore and took the license away from the pharmacist. The doctor lost his income, but that was all. They couldn't prove he knew what was going on. His partner took the fall."

"That's the shits," declared Arn.

"If I'd had any willpower, I could've done something, but I didn't figure that out until too many systems had become compromised. Now if one thing doesn't get me, another will." Merv signaled for another beer.

The TV sets around the room were being changed to the noon newscasts. When Babs brought Merv his beer, she asked, "Today is taco day. Do I have any takers? It comes with a salad."

That signaled the breakup of the Pessimists' Roundtable. Will made an excuse to leave. Everyone knew that his budget didn't allow for lunch and more than one beer. That single brew and a few minutes of conversation was his daily break from the four walls of his tiny apartment. He slowly made his way home and cooked some Ramen noodles for lunch.

Hep bailed too. His tolerance for alcohol wasn't very high. He had to get home while he was still able. If he waited until his condition showed, Babs would cut him off and he'd be stranded until one of the members drove him home, which was still an embarrassment.

Merv was paying lip service to his condition by ordering only a salad. As soon as he was alone, he'd snack big-time on all those forbidden items that tasted oh, so good.

After they'd eaten, Arn drifted to the bar to watch something on TV. Dane began his journey home, leaving Chucky and Merv at the table.

A long silence ensued as the remaining pair watched the activities at the bar. Chucky shifted his position in his chair, which drew a glance from Merv.

"Hey Merv, what do you see in your future?"

Merv swung his massive head to face Chucky. He hesitated for several beats before replying. "Absolutely nothing."

"Nothing?"

"Nothing that I want to contemplate. Oh, I already know what my future is. I'm on a downhill slide and there is no way to stop it. Like I said before, if I could put a patch on one system, there is another one waiting to fail. My only hope is that a fatal one will happen before one that would leave me with a long-term painful disability. My greatest fear is that something will happen so that I can't get to my gun."

Chucky left Merv's fear hanging in the air for a long time before saying, "I guess I'm not as bad off as I've let my self-pity tell me. My future is not pre-determined like yours. The chances are that I'll live to a ripe old, useless age. My genetics and health will probably team up to keep me in this wheelchair for a long, long time. That is my burden." "

"It can't be all that bad," quipped Merv. "You can get around better than I can."

"Before you went to Nam, what did you figure would be your life's work?"

"I planned on going back to school to get my MBA and work toward being an executive in the business world."

"That's all closed for you now?"

"Even if I could lose 400 pounds, I could never pass a corporate physical even if I didn't have a known history of drug addiction."

"Yeah, I see what you mean."

"How about you? Where were you headed?"

"My life always revolved around my physical abilities. I had to be out in the air competing. I came from a small town where it's hard to get noticed by a sports recruiter. By really pressing, I finally landed a contract with a minor league farm team as a centerfielder with a respectable batting average. Looking around me, I figured I wouldn't be there long before being called up.

"I had secondary plans of becoming a good enough golfer that I could get my PGA card because golf can be played for many years after baseball. Nowhere in my plans had I figured in a sedentary life...indoors. Yeah, everything was beginning to come together and then I got called up and sent to Bosnia Herzegovina. You know, I didn't mind Armying. It was physical and outdoors. I was having fun until this." Chucky made a hand gesture indicating his missing legs.

The conversation ceased as both men finished their beers.

"I've got to go," said Chucky.

"Me too," replied Merv, who began getting into position to hoist his bulk. Chucky waited until Merv replaced his seat along the wall. Both men moved to the door, which Merv held for Chucky.

CHAPTER 2

Tuesday was pretty much a repeat of Monday, except the Pessimists' Roundtable was noticeably more glum than usual. There were long silences, "Whats wrong, guys? Did I miss the full moon or something?" said Babs as she brought a couple of refills. She got two shrugs and a wrinkled chin for her efforts.

After Babs returned to the bar, Arn said, "Well, I know what's wrong with me. My arthritis kept me awake most of the night. What are your excuses?"

"Those stories about people doing you wrong and never paying for it brought back memories I buried years ago." Dane took another sip of his vodka. I got to comparing my life and what I have to look forward to against my ex-partner. I have some friends who know what happened and they keep me up on José's activities. He is a bit younger than I am and in better health. He sits around like a potentate with people waiting on him hand and foot. He has the best medical care. He's got a flock of kids and grandkids who bow and scrape because he controls their futures with his money. He's got that way with my money."

"Why are you worrying about it now instead of 50 years ago?" said Chucky.

"Back then, I didn't want to admit how stupid I had been. I

11

wanted to show him I could succeed on my own, but I never could."

"How'd he get to you?" said Hep.

"After I got home from Korea, I chased the buck until I had a nice little nest egg. I bought a new Ford van and headed south. I'd heard there was good money making prospects in Mexico. I batted around a bit. Eventually, I ended up in Guanajuato...in the center of the country. There were a lot of rogue weavers around there but they had lousy designs. I figured that if I could improve on the designs and control the quality, I could have a marketable item. I was good at designs and production. However, sales and distribution were my problem."

"That's where the partner came in?" said Hep.

"Yeah. I found José Matta, who was a Texas Chicano with a big family in Guanajuato. A gringo can't deal directly in Mexico. One needs a Mexican front. José was a born salesman...fluent in both English and Spanish. We got together. I'd set up the production. One of José's relatives formed the legal entity and José was the sales department.

"With my new designs, we got off to a roaring start. José brought in a stack of orders...more than I could produce without expanding. I financed a bigger building and more. I even put my new van into the business to build collateral to qualify for business loans.

"Then things started going south. Raw materials went up and labor problems developed. José couldn't negotiate contracts based on the higher prices we needed. There are some strange labor laws in Mexico. To avoid a long sad tale, the workers took over the business. I lost everything. However, it wasn't what it appeared to be. I saw José driving my van around town. When I jumped him, he said he made a deal with the workers because they could weave carpets but not sell them. That meant they were still out of work. They supposedly gave his family the business in return for work. I was set up. They even used my designs. I was really bitter over that and I lost my spark. Nothing ever went together properly after that. You can tell that by looking at me now."

"I can see why you'd be bitter thinking about it. I'd get pissed off too," said Will.

"Same kind of thing happened to me," said Arn. "Only my problem can still be a problem. I used to be a truck driver. Then I became an independent hauler with my own big rig. I went all over the states. Back in those days, I was making enough money that my accountant said it would cost me money to work more than 10 months.

"I was returning home from my last run before my wife and I were heading for Hawaii for a couple of months. I had some time to burn and got into an impromptu shuffleboard competition at a truck stop. While I was there, someone got into my truck and tore a page out of my company checkbook and stole a folder of credit cards and ID. Everything was in a briefcase. I didn't miss it until I got back from a Hawaii. By then it was too late. At that time, there weren't the credit card checks that we have today. It became apparent that the SOB who stole my stuff was also a trucker. He spent thousands and thousands of dollars all over the country with my credit cards and checks. He reprinted a batch of my checks and used them for years. He also used my ID to register in motels and then skipped.

"He must have been a nasty SOB. I've been arrested twice on battery warrants. He busted up a guy in a tavern. When the cops would trace him into a motel, my name was on the registry. He disappeared. On several occasions, down through the years, I've been arrested on his warrants. My credit never did recover."

"Did you ever find that guy?" asked Dane.

"He became famous. The police nicknamed him 'The Trucker'. They never did catch him but by accident I recently found out who he is."

"Did you turn him in?" said Dane.

"I thought about it but too much time has passed and it would cost me too much to pursue the whole thing. It wouldn't make any difference in my life now."

"How did you identify him?" asked Chucky.

"Before it became too painful to walk the beaches, I used to watch the Snowbirds surf fish. I got to know several of them. I ran into one guy from Des Moines. Once we got to talking about nicknames. When he heard my real name, Arnold Leitner, he said he used to have a neighbor by that same name. This guy didn't like 'Arn', as he was called. We got to comparing notes. This Arn was a retired truck driver...a real bastard. He had lots of money and for a pastime he ran a booth at the local flea market. He always seemed to have a lot of new stuff. Everyone thought he was a fence.

"Anyway, I'm sure I have the right guy. Too much matches up. Then he used my identity to set up a new life in Des Moines... even got married under my name."

"Why haven't you put a bullet between his eyes?" demanded Hep.

"Inertia," said Arn, as he tipped his beer can back far enough to get the last drop.

"Inertia?" said Chucky.

"Yeah, it takes too much energy to get an old sedentary body like this into motion. Also, I figured I had too much to lose. With my luck, as soon as I plugged him, I'd turn around and bump right into a policeman." Arn shook his can and it was still empty. He set his can down on the table a little harder than he'd anticipated. It thumped loudly. "I guess there isn't much to lose anymore."

"Hep, do you have anyone you'd like to plug between the eyes?" asked Arn.

Hep bobbed his head. "You bet. That suggestion was one that I considered for a long time. My wife, Alice, had a bad pregnancy. She lost our son and they had to give her a hysterectomy. It was touch and go for both of us, but we eventually accepted the fact and we used to joke that we were stuck with one another and we'd just have to grow old together. It was all working out pretty good. When I retired we started working on a long list of things we'd never been able to do while I was employed. We were having a ball.

"We were traveling in Oregon. We were headed back to our motel late one night when an oncoming car turned in front of us. Alice was badly injured. An ambulance took us to the local hospital. They didn't have a doctor on duty in emergency. They had one on call. It took him an eternity to get there. They kept me in the waiting room. I was sitting so that I could see into the emergency room whenever anyone went through the door. Alice was just lying on the table.

"Finally, a guy in civvies' went in. Later I saw Alice on the table naked with a rib expander in her chest. The guy had his hands inside her working on her heart.

"A while later, one of the nurses came out to tell me that Alice had died. While she was talking to me, the guy staggered out of the door and down the hall. That SOB was drunk. I chased after him and caught him just as he was going through a door. I spun him around. He reeked of alcohol. His eyes were bleary. He mumbles he was sorry before lunging through the door. A couple of nurses grabbed me. I heard the door lock and he was gone.

"I shouted that the doctor was drunk. The nurses said he was just distraught over losing a patient. I know a drunk when I see one. I used to manage large nightclubs. The bastard killed my wife."

Tears welled up in Hep's eyes. He batted them away before saying, "She left me all alone."

The rest of the table turned their attention to their drinks as they dealt with their own feelings.

"Is he still out there?" said Chucky.

"Yes. I tried to get his license but no one would substantiate my claim. It's really hard to get any medical workers to accuse a doctor of wrongdoing. I couldn't get anywhere, especially since I lived thousands of miles away."

"I wouldn't be inclined to put a bullet between that drunk's eyes," said Chucky.

All eyes swung his way. Everyone had the same question.

"You suddenly opposed to killin'?" demanded Arn.

"No. Not at all." said Chucky. "However, legally whatever crime the doctor may have committed was against Hep's wife, but who is suffering? Hep. My feeling is that retribution might be administered more appropriately if the bullet entered the skull of someone that the doctor really cherishes. The retribution is that the doctor knows that he was the cause of a loved one's death. Let him live with that as Hep lives with the results of the doctor's action."

"Yeah," said Hep, "If I put a bullet in the doctor's head, he wouldn't suffer as I have. He'd be getting off easy."

Dane nodded agreement. "Killing José wouldn't really satisfy me either. I'd rather he had to live like me...alone and broke. However, I wouldn't wish this emphysema on even José."

Dane struggled to his feet to begin the journey home as Will asked Chucky, "Don't you have someone on whom you would do like to heap some hurt?"

"Oh, yes."

"Direct action or let him suffer a while?" asked Hep.

"This one would be direct action...after he knows what it's for. I'd like to kill him twice...once for each leg."

Dane sat back down. "Don't stop there."

Chucky looked off into the distance for a bit. "I try not to think about this because it makes me so mad nothing else seems important. The whole world can go to hell until I calm down again."

"Are you lethal when you're upset?" quipped Hep.

"Not to you," said Chucky. The way that it was said left everyone wondering what had not been said.

Everyone waited.

Chucky took a deep breath and adjusted his position, in his

chair. "It's a very short story. It all happened in a flash. A bunch of us were asleep in a shelled out stone farmhouse while we were pushing into Bosnia Herzegovina. I woke up to an unnatural sound. It was just beginning to get light. The sound was coming from our guard, Corporal George Wesley Holmes, who was giving a frightened animal squawk. He had his rifle pointed at the chest of a rag head who was standing in the doorway. Both were frozen. Then Holmes dropped his rifle, screamed like a little girl as he dove behind a broken wall. Before I could bring my weapon to bear, that bastard, who should have been dead, tossed a grenade and dove out of sight. Two guys were killed and I got this." He made a motion to where his legs should have been. "They were the lucky ones."

"What happened to Holmes?" said Will.

"After I recovered enough, I told my story. I figured the army would take care of Holmes. I had other things to worry about. It was weeks later, after I was shipped from Germany back to the States, an investigator came around asking questions. He said Holmes had reported that he had been on guard duty earlier, but he turned the watch over to me and had gone to sleep behind that wall. That was why he didn't have a scratch. He didn't know anything about it until the grenade went off."

"That dirty coward!" said Dane, with passion. "Did he make it through the campaign?"

"Yeah." I've located him on the Internet. He's a minor executive in what appears to be a family business. At least it has the same name."

"Where?"

"Venice, Florida."

"That's not far," said Will.

"It is if you're pushing a wheelchair," cracked Chucky.

"Tell me about it," wheezed Dane as he struggled to get up again to begin his journey to the car.

CHAPTER 3

When Chucky worked his way into the post, Hep was already at the table. It appeared he been there for some time. There was a pall of smoke in the corner and the ashtray was full. When Babs brought Chucky's beer, Hep slid the ashtray across for her to pick up. The smokers normally didn't smoke at that table because of Dane's emphysema.

Hep was already past his normal state of inebriation for that time of day.

As Chucky adjusted his position to the table, he said, "What's biting you today?"

Hep focused his bleary eyes on Chucky. "Our table talk the last few days."

"How's that?"

"We're all a bunch of bloody losers. Everyone of us had someone do us a major wrong and they got away with it and all of us were too chicken shit to do anything about it. Last night, while I was taking all those pills that keep me alive, I was looking in a mirror. There isn't much time left to clear the slate."

"Does it need to be cleared?"

"Damned right it does. I should have done it decades ago, but I was afraid."

"Afraid?"

"Yeah. I was afraid that if I did something to that bastard, I'd get caught and lose whatever I had. I was afraid of everything. If I did something bad I'd be shunned by friends and family. A few days ago, I was blowing off about 'what did I have to lose?'. Well, I must have felt I had something because I still wasn't ready to follow my own pronouncements.

"If I were to do something illegal and get caught, all I'd miss is this." Hep held up his beer. "And you guys."

"What did you decide?"

"That's where the rub comes in. When I do something to him, what do I do? We talk that the retribution should be appropriate. Killing him would be too good for that bastard. But to do something else takes time and money."

"What do you mean, money?"

"I know who he is and where he is and it wouldn't take much doing to hit him, especially if you don't particularly care about getting caught. But to do something to someone that will hurt your target requires research...maybe a lot."

"I see what you mean," said Chucky. "It wouldn't do your cause much good to hit his wife if she was suing for divorce and wanting the house and half of his money. Perhaps, that would be doing him a favor."

Further discussion was delayed as Will showed up early.

"You guys are out of sync," said Babs. "What's up?"

"Didn't want to risk missing you loading the beer cooler," said Will.

"Boy, you guys must be hard up," said Babs, as she stuck her hand out for the money for Will's beer.

It would be exact change +10% tip. She glanced at the other two at the table and regretted her smart comment about being "hard up."

The rest of the gang straggled in. Hep spread his mood over everyone.

"Okay, Hep, what's chewing on your tail?" said Arn.

"After our talk yesterday, I've been staggering around feeling completely useless. I'm coming to the end of a long downhill trip. There is nothing for me in the future. I'm not even a candidate anymore for the harvest of body parts."

"I know what you mean," said Merv. "When you know that to-morrow is not going to be any better than today, why bother?"

The mood around the table continued to degrade with Hep's advancing drunkenness and wild, disjointed oratory.

Babs caught Chucky's eye, then made a throat cutting motion and nodded at Hep. She always hated denying him service be-cause of the fuss that always ensued. She was passing the buck to Chucky.

"Hey, Dane, can you drop Hep off at his apartment?" said Chucky.

"If he can make it to the car. I can't pick him up."

The slurred retort, "I can make it," didn't engender much faith.

"I'll go with them and see that Hep makes it," said Merv.

That was enough to start a major exodus, leaving Chucky alone at the table. He finished his beer and gave Babs signals for lunch, which would be a hamburger that day and a cup of cof-fee. A fuzzy little idea was flitting around in his mind. He needed some time to snare it.

To reduce social intercourse in the club, Chucky pulled a legal pad from his wheelchair saddlebags giving the appearance of writing a letter. Actually, he was taking notes as he tried to ar-range his thoughts.

The next morning, Chucky was back at the Legion Hall before any of the gang arrived. He needed a little time to check his thinking and polish his presentation.

CHAPTER 4

As the pessimists of the roundtable filtered in, Chucky was measuring attitudes and frustration levels. No one was quite as stressed as on the day before, but agitation was still running high.

After the group settled down, Chucky scooted his chair up so he could place his elbows on the table. The action attracted attention. Chucky leaned forward and in a conspiratorial tone, he whispered, "How about one more mission...a last hurrah?"

No one could even guess what Chucky had in mind, so no one said anything. They just waited expectantly.

"Yesterday, all of us were feeling pretty sorry for ourselves. We were complaining about being useless with no futures worth contemplating. At the same time each of us was remembering an old wrong done to us and lamenting the fact that we'd never righted that wrong.

"Let me ask you a question. Why didn't you retaliate against those who did you dirt?"

There was silence around the table as each man examined the secret side of his life.

Arn finally said, "I kept telling myself and anyone who wanted to know the story that I could do it again. No sweat. But the real reason was that I was afraid to try to take on a Mexican in Mexico. Gringos don't fare well in the Mexican system of justice. If you cross the wrong person down there, you can wind up in jail or dead. I was afraid that anything that I did would just make the situation worse and destroy any future I might have."

Arn lapsed into a sullen silence.

"What have you to lose now?" said Chucky.

"Not a damn thing."

"What would you like to do to the joker?" said Merv.

"Down through the years, I contemplated various paybacks, but I've never really come up with a satisfactory answer. Killing him wouldn't be that satisfying. I'd rather he had to continue to live, but in a state of poverty. It's his turn to be miserable. Of course, he's almost as old as I am so he wouldn't be miserable all that long."

"You could probably devise some operation where his misery is passed on to others he holds dear," said Chucky.

Chucky was aware that Dane was pulling his air in through his nose and blowing it out through pursed lips. He was building his oxygen supply. He was preparing for some exertion. Since it was too early to go home, he was probably getting ready to express his opinion, which sometimes could be rather lengthy.

When he was ready, Dane said, "Okay, Chucky, what's all this about? You've given enough of a pep talk. Now it's time to tell us what type of game you want us to play."

Chucky leaned back in his wheelchair. "The conversation of the last couple of days has shown that all of us have unresolved problems and all we do is bitch about them. We can't resolve them, so they continue to chew on us. I have decided to do something about mine." Chucky reached into a pouch on his chair for a sheaf of papers and stacked them on the table.

"There are two ways of going about this. I can do it myself or I can get somebody to do it for me. If I have somebody else do the job, there will be a price...unless...I trade problems with somebody else."

"Why would you want to trade off with somebody? That would reduce the satisfaction." said Dane.

"I'm not so sure I can be objective. I'm too close to the problem and my hatred could cloud my judgment and interfere with my performance at the wrong moment. Is this a kamikaze attack where I blow my brains out? I have to be prepared for that before I start. I can't count on the police or anyone else killing me.

"If on the other hand, I want to get away with it, I'll be at the top of the suspect list and a double amputee in a wheelchair is hardly able to fade into the woodwork. Also my name and his name are inexorably tied to each other due to our cross accusations.

"Right now it doesn't really make any difference whether I am caught or not. If I'm captured, I'll eat three times a day without crawling around the kitchen cooking and washing dishes. I don't have to worry about clothes, medicine and the list goes on.

"Granted, I can't go anywhere, but where do I go now? I wouldn't be able to come here for a beer or two. That would probably be my greatest loss. No, I'd give this up to have someone take care of me for as long as I last."

Chucky stopped talking as Babs headed around the end of the bar to bring the refills that had been ordered by sign language.

After Babs left, Chucky passed out two sheets of paper to each man from the stack he'd placed in front of himself. "I've written up a short synopsis of the events where I lost my legs and a summary of the various testimonies that have been given. The second sheet contains all of the information I have on the bastard's current situation...address, employer, family information. A lot more can be found on the Internet.

"I'll trade my vengeance, justice, retribution or whatever you want to call it, with anyone at this table. The deal is, that I'll provide you with all the information I have. It will be up to you to judge the severity of the wrongdoing without any consideration of my feelings and think up a just retribution and execute it. I will accept your judgment.

"In return, I shall do for you what you are doing for me.

"There should be a time limit. I should think that two months is an adequate amount of time to put my affairs in order, familiarize myself with the target and situation, travel to the site and execute the mission. Do I have any takers?"

"Wow," said Dane.

The rest at the table shuffled uneasily.

Hep pinched in the middle of his beer can until it folded over. "You're too young to do something like this."

"If you were my age and had to look at the rest of your life in a wheelchair, you might see things differently. I'm just smart enough to know that no great theories of time or black holes are going to be coming from between my ears like that Britt in a wheelchair. I'm a grunt...a grinder with limited intellectual skills. Reading bores me to death and television strains my senses of reality. When having a beer or two with you guys is the highlight of my day, it's time to do something different. At least if I can pull off the mission successfully, I'll have contributed to a worthy cause."

Chucky pushed himself away from the table. "Think about it. If anyone is interested, tell me tomorrow." Chucky stopped by the bar long enough to pay his tab before wheeling himself out to a specially equipped van.

CHAPTER 5

The whole Pessimists' Roundtable arrived within five minutes of one another. Conversation was restricted to inane comments until everyone was served and settled. A silence fell over the table. Chucky figured that the rest were waiting for him to ask if there were any takers for his proposal. He was surprised when Dane took the floor.

"Chucky, after you left yesterday, we hung around a while longer. We came to a consensus. We're all in. We'll exchange grievances and complete our assigned missions. On the personal side, I already feel purpose coming back into my life. Have all you guys written down your info?"

Each pulled what looked like a folded letter from pockets and passed them Dane.

"Chucky, give me yours."

From a wheelchair pouch, Chucky produced a sheaf of papers from the preceding day. He peeled off two sheets and folded them into the proper form.

Dane had six number 10 white envelopes ready. When the info sheets were in the envelopes, Dane shoved them into the center of the table and shuffled them thoroughly.

"I've made up some rules. Each of us will take an envelope. We'll go around the table to open them one at a time. Should you draw your own, we do the whole thing again until our missions are properly distributed, and the two months begin."

"Oh yes, Chucky, when we're talking this out, we decided to extend financial and material assistance to anyone who needed it. Some missions will require more travel than others. If Will draws one a long ways away and he needs ticket money, we'll provide it."

The group watched Chucky intently as they tried to judge his reaction to their decisions.

When Dane signaled he had finished by sitting back in his chair and picking up his beer, Chucky let a small smile rippled across his face. "Now, we'll show what A Few Good Old Man can do. I'm glad everyone's aboard. This should be an interesting couple of months."

"Let's see what we have to do," said Merv.

"Chucky, you go first. Check inside the envelope to make sure it is not yours. We'll read them once everyone has an envelope."

It took three tries before everyone was set. Chucky ordered another round before they started reading.

"What are you goofballs up to?" said Babs as she brought the beers.

"Planning for a secret mission," said Arn. Everybody chuckled.

After the barmaid was out of earshot, Chucky said, "Babs is getting really curious. She's beginning to hover when she has an excuse to pass by. Let's keep her guessing. What we're proposing isn't protected by any law that I know of."

After a conspiratorial chuckle everyone opened his envelope and read the contents.

Chucky scanned down his assignment. It was Arn's identity thief. He was satisfied. That should be interesting.

Before Chucky got too deeply involved, he decided to do some

record-keeping. He pulled out a sheet of paper and drew a verti-
cal line to make two columns. He put his name in the left and
Arn's in the right before passing it around the table. When the
sheet returned to him, he surveyed the completed list before
passing it around again so all the guys knew the assignments.

When the list returned, Chucky made brief notations as to
what the injustice had been.

Chucky	Arn	ID Theft
Arn	Will	Bad Drugs-grandson
Hep	Dane	Mex. Rugs
Merv	Chucky	Coward
Will	Merv	Dr./Old Drugs
Dane	Hep	Drunk Dr.

Chucky maneuvered himself around to minimize his discom-
fort before saying, "Arn, I need to spend enough time with you to
get more information on this joker and his family so I can make
a decision on what I want to do and how I'm going to do it. How
about coming over to my apartment? Then I'll have access to my
computer."

After an appointment was set up, Chucky turned his attention
to the general group. "If we have any valid addresses for all these
guys I may be able to provide you with a close-up aerial view of
your client's residence. In some cases Google may have photos
of the front of the houses. Other info might be available on the
Internet.

"Merv, you're computer savvy. You can help too, right?"

"Sure. Give me a call most any time. I have a speaker phone
so I can listen, talk and type at the same time. Anything that has
to be printed out, I'll bring in the next morning."

"Good," said Chucky. "These one-on-one sessions shouldn't
be here. That would vary too much from our usual routine. It

might raise some questions we wouldn't want to answer."

"Today's hamburger day. How about staying for lunch and then I'll stop by your place?" said Arn.

"That sounds good to me," said Chucky.

CHAPTER 6

Arn followed Chucky's wheelchair-modified van to his home. Chucky could afford better accommodations but his tiny apartment suited him. The landlord had divided an older house into three small units. The made-over garage was perfect for wheelchair access. There were no steps. Only a slight incline and a threshold had to be negotiated. All the walls were concrete block and the landlord had given him permission to anchor bars anywhere. A small kitchen had been built in the left rear of the garage.

There had been a step up into what had been the master bedroom with its bath and walk-in closet. A carpeted ramp had been installed. The landlord was a fine handyman, who could put in anything Chucky requested.

The interior of Chucky's apartment looked more like a jungle gym than living quarters. When he was home, Chucky forsook his wheelchair for a four-wheeled dolly. With the steel bars he could get in and out of his wheelchair and his lounger more easily. He could maneuver around the house better on the small platform. All the living room walls were lined with 2 x 12 planks atop single drawer filing cabinets. The working surfaces were just the right height for Chucky when sitting on his dolly.

Chucky's computer equipment took up much of the surface space.

Arn had never been to Chucky's apartment before. "I'd never given a thought to what equipment you must have to live. Fascinating."

Chucky moved under the bar that he used to hoist himself from the wheelchair and lowered himself onto the dolly.

"I want to work on the computer. You can sit in the wheelchair. The recliner is too far away.

While Arn was getting positioned, Chucky scooted away for a couple of beers.

After a prolonged guzzle, Chucky opened a new file. "What's this joker's real name?"

"I've never found out. Everything is in my name. To the police, he's known as 'The Trucker'."

"Have you ever passed this information along?"

"No, I can't prove it and it's too late for me to start an investigation."

"Let's see what Google has to say about 'The Trucker'." It only took Google a fraction of a second to locate over 19,000 entries. "Of course, the rock 'n roll singers received the first billing but it looked as if there was plenty of material on our boy. I'll sort through this later."

For the next hour, Chucky gathered the personal information from Arn that the thief could still be using. He printed maps and aerial views of the house. The police had complaints against the subject, but nothing that led to an arrest where fingerprints might have been taken.

Chucky had ample avenues of inquiry that he could pursue. So he turned his attention to Arn's target.

"Will says that Mo's house has been foreclosed on," said Arn. "He doesn't know where the kid lives, but he's still in the area."

"It shouldn't be too hard to find him," said Chucky. A Google search of the kid's old address of 140 NE 1st St. showed where the house was located. "Will says that this kid, Mo, was always in trouble with the police. This address is out in the county. Let's see what we can find on the sheriff's website. Oh, looky here. The sheriff seems to have established a long-standing re-lationship with Mo."

Chucky went down the chronological list. His search produced the 140 address. It was followed by a couple of entries from an apartment on the Beach Highway. The last three are 141 NE 2nd St." Checking the map, Arn laughed. "His new address backs up against his old house. He must like the neighbor-hood."

"It's more likely he wants his clients to know where he is. Didn't Will say something about his customers walking down the easement behind the house?"

"Yeah, now they just have to turn right instead of left. I guess I'll wander over there and take a look."

"Anything else you want off the computer?"

Arn shook his head. "Naugh. All I need to do is identify Mo and learn his schedule."

Arn started to leave but he was trapped in a wheelchair. He needed to lever himself onto his feet, which pushed the chair away. He floundered around until Chucky set the brakes.

"Thanks," said Arn as he tried to cover his embarrassment at not being able to perform the simple act of getting up.

CHAPTER 7

After dragging himself from the aging Chevy Malibu, Arn patted the fading hood paint. "I've been hoping we could finish this enlistment together. We may just make it."

Arn dumped his notebook on the cluttered dining room table before heading for the bathroom. The long vigil of watching Mo's activities added substantially to the urgency of the trip. Afterwards, he made a stop at the refrigerator for a beer before sitting down to enter the information he'd collected into his charts. His intent was to watch the subject until a predictable routine was established. Arn had tentatively allotted a week to the chore, but after just one afternoon certain other possibilities had presented themselves.

Will had been right. That 15-year-old was a real hulk. He was about six-foot tall and probably carried a couple of hundred pounds. He could handle that much weight on his big, broad frame. Arn figured that when Mo gets a little older, he'd turn to flab if he doesn't take care not to.

The next day, Arn was the last to arrive at the club, which surprised him. Usually the guys wandered in somewhere around the appointed time. Everyone seemed more alert and animated. The various topics of conversation did not revolve around their impending missions. Everyone was enervated.

That was notably different energy. "What are you old buzzards up to? I'll bet it's bad news for someone."

"Our Viagra dealer's dropping by today," quipped Hep. "When do you get off duty?"

"You'll need a double dose," laughed Babs.

Outside of making some appointments for computer inquiries, nothing concerning the missions was discussed.

After the pessimists broke up, Arn stopped by a fast food place for a couple of burgers and coffee. He took up a position where he could monitor the traffic in and out of the easement behind Mo's house. When he arrived at 12.30 p.m. there seemed to be no activity. Arn was beginning to believe that Mo was not home, when at 1:00pm a white girl pulled into the parking lot of a bar across the street. She was too young to go into the bar. She looked about before striking out for the easement.

It wasn't until 1:45 that a disheveled girl returned to her car.

At two o'clock, activities picked up. There was a pretty steady flow of white and Latin young people. Only one black dropped in. Arn had the feeling that that one guy might be a supplier.

Arn watched the house for two more days. The following morning, Arn was the life of the group. He made a point of engaging everyone in at least a brief conversation. He was also the first to leave. He walked by Chucky's wheelchair and dropped a key on the table and said "Keep the house safe from tigers."

Everyone was quizzically watching Arn slowly make his way to his old car.

"What was that all about?" said Will.

"That was Arn saying goodbye," said Chucky. "He's on his mission. Pay attention to local news."

In his car, Arn took a swig of bottled water to wash a couple of Advil down before heading out to the parking lot. Before pulling onto the street, he hesitated a couple of beats, taking a last look at the Legion building.

One of Mo's patterns seemed to be that between noon and two he closed the operation down. Sometimes he went out to eat...so his car would be gone. Most of the time he had a female visitor who normally arrived precisely at one and left 30 to 45 minutes later.

At 1:48 a cute girl in a supermarket uniform hurried to a little green car, clutching a small purse to her chest. As soon as the girl was out of sight, Arn climbed out of his car, carrying a cloth shopping bag from the same store as the girl's uniform.

There was a well beaten path between the two houses facing east. The wire on the end of a 4 foot chain-link fence that had closed off the yard from the utility easement was rolled back, leaving an 8-foot gap.

The backyard was a dried up old weed patch. Arn knew that the back door was never locked. Customers just walked in.

Arn also knew the layout of the house because he had a real estate agent show him the same design a couple of blocks away. The back door entered into the utility area off of the kitchen. Beyond the kitchen was a dining area. The living room went to the front of the house. The master bedroom and bath were off the living room to the left.

Arn wasn't afraid of being spotted on his approach, since all of the windows were blanked out so no one could see in.

By the time Arn, reached the back door, he was fully committed to his mission. He didn't hesitate at the door other than to maneuver the bag onto his left wrist so he could use both arthritic hands to grip the doorknob.

Arn eased the door open a couple inches. As in most older homes in hurricane-prone areas, the doors opened outward. He felt a little resistance. In the upper corner was a simple, home-made warning system. A large nail was driven into the side of the door jamb. Affixed to the top of door was an "L" shaped strip of spring steel. When the door was either open or closed, the spring steel scraped across the nail. When the strip was released, it gave off a metallic twang.

To keep the alarm silent, Arn placed a hand on it as he opened and closed the door. There was a TV playing in the living room. Arn slipped along in his sneakers. When he stepped free of the kitchen, he found Mo stretched out in an oversized recliner, stark naked. He was playing with himself. The recliner was positioned next to the door into the master bedroom.

Arn's movements caught Mo's attention. "What the fuck are you doing in here old man? Get out."

Apparently, an old, arthritic man didn't represent any particular hazard, even when Arn didn't scurry out. Instead, he moved to the foot of the recliner so that he was looking at the soles of Mo's feet.

"A guy said I could get some stuff here. He said you might trade." Arn had one hand in his bag gripping an ancient Army colt .38 revolver that he had had since he was a kid.

Arn's words had momentarily halted Mo's motion to lever himself upright with the handle on the side of the chair.

The shopping bag had been carried waist high. When the straps were released, it fell away. Arn gripped the revolver in both hands and in one motion brought it to bear and fired. The hammer had been cocked outside.

Mo started to let out a yell when he saw the gun but it changed into a scream of anguish as the copper-washed, soft nosed .38 slug smashed into his groin, shattering the pelvis as it plowed into the lower body cavity.

Mo was whimpering. "Help me man. I hurt bad," he pleaded. When no offer of assistance was forthcoming, Mo whispered, "Why?"

Arn didn't answer immediately because his teeth were clenched against the pain. He moved to the davenport so he could sit on the arm. The gun dangled from a finger. Arn stuffed the revolver into the waistband. He'd expected to have to shoot at least twice, but maybe he could get by with just one. Mo was trying to stay immobile because movement was too painful. He started crying and lapsed into begging. "Please, please, it hurts so bad. I never did nothin' to you."

Once Arn could speak, he snarled, "No you didn't do anything to me directly, but your bad drugs killed my friend's grandson. This is for the boy and his grandfather."

Arn glanced at his watch. He was coming up time for the dope store to open. So far there was no indication that anyone had heard the shot and raised the alarm. Arn made his way to the back door and locked it. He didn't want anyone else to come to his party.

He returned to the living room. Mo was still holding his crotch but he couldn't stem the flow of blood and body fluids. A good-sized puddle was forming on the floor.

Mo's face was contorted with pain. He'd ceased talking and was only moaning. It wouldn't be long now. Arn sat down on the couch for his death vigil. He started twice when somebody rattled the back door. As time passed, Arn searched around in the corners of his mind to see if there was any compassion for Mo. He couldn't find any. Oh, there were feelings about the conditions that produce the Mos of the world, but even 15-year-olds know when they have crossed the line.

Using his elbows, Arn levered himself to his feet. Mo's eyes were only slits. The active bleeding had stopped. Arnold felt the neck for arterial activity and found none.

Arn made his way to the wall telephone over-the-counter between the kitchen and dining room. He had to lay the phone on the surface to punch the 911. His hands and arms hurt miserably.

When the operator asked the nature of the emergency, Arn said, "I just shot and killed the occupants of this house. Please notify the sheriff's office. There is no big rush. The dead guy is not going anywhere and neither am I. I'll open the front door."

Before the dispatcher could ask any further questions, Arn dropped the phone on the countertop. He left the line open as he shuffled off to unlock the front door.

As Arn turned away from the entry, the heavy revolver almost fell from his belt. Talking to himself, he said, "I better not be carrying this when the cops arrived." Arn put the gun on the coffee table, which was in sight of the front door.

The adrenaline started to fade along with Arn's energy. He needed to sit down. He rejected the davenport because it gave him a crotch view of Mo. The gun was too close and he'd have to eventually struggle to get up. He ended up in a straight back chair at the dining room table.

He wished he'd turned off the TV, but he wasn't going to expend the energy now. His eyes moved over the scene. Mo must have had sex in the recliner. His clothes were in a pile on the floor. They were soaking up the blood. He didn't see any signs of Mo's guns. If they were small caliber, they may be in his pockets.

The house was a mess. The large flat screen TV, the recliner, the audio system and his clothes all spoke of money, but the house was a pig sty. There was nothing in the house that revealed any personality or special interest other than a couple of large rap posters taped to the wall with bits of duct tape. The kitchen was a disaster area. The sink was full of dirty dishes. The garbage can was overflowing. Arn became aware of the stench of rotting food.

A siren started to tickle the fringe edge of his hearing. It wouldn't be long now. Arn had figured that anxiety would upset him when what he had done became public knowledge. However, Arn found himself relaxing despite the continuing pain in his hands. He had to hold his hands to keep them from shaking, causing even more hurt.

Instead of being anxious, Arn had a sense of euphoria. He had succeeded in his mission. Now he could get rid of all of his self-incrimination for not having taken action years before. The wheels of retribution were in motion. He was confident that Chucky would handle the situation. Now he could settle back into a future that was pretty well set.

Arn's gaze fell on Mo. He still couldn't muster much sympathy for him even though he was just a kid. The parents would probably be outraged in public and relieved in private.

The deputy turned off the siren. Arn imagined there was also a backup car in route. Arn wondered if the 911 operator was still on the line but he wasn't curious enough to go find out.

Arn smiled. For an old duffer, his ears were still pretty good. He could hear the sounds of the car doors opening. It wasn't right in front of the house. The chain-link gate clicked, which meant somebody would be in the backyard. A third car arrived. It was not trying to be quiet. When a little sound came from the vicinity of the front door, Arn yelled "Come on in. Just shove the door open and you can see me sitting at the dining room table."

"Come out with your hands in plain view."

"You'll have to come in. I'm an old man and I don't have the energy to do so. I'll put my hands on the table."

After a couple/three exchanges that didn't bring Arn out of the house, the door was swung back against the outer wall. A shotgun toting deputy peered into the room. In stages, five deputies eventually populated the living room. Arn was searched and relocated to the garage under guard. No one had much to say to him. Questioning was being left to someone else.

Ultimately, Arn ended up in a small room in the sheriff's office. There was a one-way window in the wall. Shortly, he was joined by a lieutenant who introduced himself and another officer.

"My name is Lieutenant Pat Lewis and that is Officer Price."

Lewis sat in the chair across the table from Arn. Price pull up a chair against the wall indicating he was just there as a witness.

"Are you Arnold Leonard Leitner?" said Lewis as he read from Arn's driver's license.

"Yes."

"Is all the other information on this license correct?"

"Yes, except the height. I'm getting shorter," said Arn with a little chuckle.

The lieutenant read Arn his rights before saying, "You shot

that kid and then called us, right?"

"Yes."

"I'm going to turn on a recorder and why don't you tell me all about it."

"Fine."

The lieutenant ran through the preamble of time and those present and all the rest of the necessary details.

Arn was asked to state his full name and address.

Okay Mr. Leitner, tell me about this affair."

Arn adjusted his position to ease his aching joints. "This is purely a business deal."

"A business deal? How so? Were you paid to get him?"

"Oh no, it's nothing like that. Look at my hands. I can hardly pick up a beer can. It's getting harder and harder to tend myself. Cooking, washing dishes, laundry...all those things that you have to do while living alone.

"I decided I'd work out a trade. This kid, Mo, is a first class crud. He would be spending a good portion of his life locked up somewhere at a high public cost. I'm just trading a long-term liability for a short-term liability. I don't have that much time ahead of me. I probably won't live long enough to get through a trial...unless you get a lot faster than you are now.

"I'm also running out of money. With the inflation, I'm having difficulty buying my meds and making ends meet. So I'll let the government feed and clothe me and provide me with shelter and medicine.

"All you have to do is get rid of the body and your liability is over as far as Mo is concerned."

"Why did you pick Mo?" said Lewis.

"I heard he was responsible for the death of a guy's grandson with these drugs, then nothing was done because Mo was a Juvie. I've been watching him for a few days and he's dealing in drugs. I imagine the autopsy will show he had sex just before he died. There was a scrawny, little white girl coming out just before I went in. Noon to 2 o'clock is when he had sex. At two o'clock business started up again."

Interrogation went on for another hour. Then Arn was transported to jail where he started his new life.

CHAPTER 8

The Pessimist Roundtable arrived promptly at their appointed time. All were energized. The reports of Arn shooting Mo had hit the local evening newscasts. By the late news, rumors were floating concerning the motives involved. The "business arrangement" was choice material for unbridled speculation.

There is a good turnout at the legion. As of yet no connection had been established between the murder and the vets' organization but the legionnaires recognized Arn.

Babs offered her condolences to the roundtable since Arn had been one of theirs. The group thanked her for her concern and smiled at her poorly concealed curiosity. She was certain that there was more to it than what appeared on the surface.

Will was the one with a broad smile on his face. After Babs returned to her duties and various legionnaires had offered their feelings on the affair, Will said, "I had reconciled myself to the proposition that I'd never live to see any retribution. I'll still cry for my grandson but I won't be grinding my guts out that his killer is still walking free.

"It sounds as if Arn constructed a good plan and executed it perfectly. Here's to a successful mission." Will raised his beer in salute. The rest of the group followed his lead.

The next day Arn appeared in court and the charges were first degree, premeditated murder, which was what he was seeking. The roundtable had mixed feelings. They were energized by the completion of the mission and that Arn had succeeded in his business transaction. At the same time they were saddened that Arn had had to take such a path. Old age, ill health and poverty are cruel companions.

"It looks as if I'm going to be next," said Hep. I have my plane ticket to Texas for tomorrow afternoon...from Orlando to San Antonio. Can anyone give me a ride to Orlando?"

"Yeah," said Merv. "I can take you. When do you want to leave?"

The scheduling had to wait until the rest of the round table offered their best wishes for a successful mission. Dane was fighting back high emotions. The feeling around the table was that Hep was not planning on returning.

"We better break this up. Babs is getting curious again," said Chucky.

"Pick me up at noon and I'll buy you lunch. I don't have to leave until 4:30."

****

Hep could see Merv coming. There was no missing that starboard list to his car. Hep's emaciated body and small carry-on bag didn't do much to right the imbalance.

The ashtray was open and stuffed with receipts. Hep leaned over to drop a couple of keys on top of the slips of paper. Merv assumed they were his apartment and car keys, but he didn't make any comment.

Merv knew quite a bit about Hep's mission since he had done the computer searches and had downloaded the aerial maps of the target's residence. It was located outside of Laredo, Texas,

on what appeared to be a ranch. The house looked rather exten-
sive with numerous outbuildings and a large barn.

"How are you going to get to Laredo?"

Hep shrugged. "Probably take a bus and then rent a car."

Neither man seemed inclined to discuss the mission either
while traveling or over lunch. Merv dropped the old man off at
the terminal and watch for a short bit as Hep slowly made his
way in the door. He wondered what type of plan could be executed with such limited physical capabilities.

CHAPTER 9

Hep needed help with the ticket dispenser. It had been years since his last flight. Then he needed some additional help to order a wheelchair for the plane change at Dallas. Even with the delays there was still time to hunt down a cold beer. He and Merv had missed their beers at the roundtable. He steeled his will to have only one. He needed all of his wits and physical capabilities until he could make it to Laredo.

The flight was uneventful. Hep refused to speculate on a mission ahead because he needed additional information before he could proceed. His main problem was getting comfortable. Nearly 6 hours in one spot was a long time for Hep. His legs kept cramping.

Finally, the wheels touched down in San Antonio. With a wry smile on his face, Hep hobbled into the expensive airport hotel. To pay for his room, he slid a credit card across the desk, which had an indecently high limit for a man of his means. He seldom used the card. He always had to know how he could pay for it before the card came out of his wallet.

He wasn't dealing in any nefarious way. If he wasn't around, his old paid-up G.I. insurance would cover any balance. The bank just might have to wait for a bit.

"How long will you be staying with us?" asked the pretty young clerk.

"Just tonight. It's too late to make it out to the ranch. I'll worry about that tomorrow."

Hep didn't make any pretense of getting an early start. The flight had been trying, so he arose when he was ready, cleaned up and had a leisurely breakfast. He had plenty of time before catching the bus south to Laredo.

Although he had never been in Laredo, he'd been in enough border towns to know the feel. After renting a nondescript car, Hep cruised around until he found a motel that satisfied his general requirements. He checked in and tossed his bag onto a bed before adjourning to the bar.

For the next three days, Hep became familiar with José Matta's ranch and movements. José seemed to be an honored patriarch with the title of "Patron." His ranch apparently backed up to the Mexican border. The main residence had the appearance of a modernized hacienda. It appeared to house the extended family. A wing to the rear looked as if there had been repeated add-ons. There was evidence of kids all over the place. Another large building turned out to be an enormous communal garage.

From the aerial photos that Merv had provided, there seem to be a large barn. This proved to be a warehouse.

Inquiries had provided the story that José Matta had moved from central Mexico some years back to take advantage of certain tax benefits given to border enterprises. He had prospered. The warehouse behind the residence was full of rugs.

Hep felt that he had collected enough information to start formulating a plan of attack. When he returned to the motel, he poured himself a scotch out of his private stock and stretched out on the bed to scheme. Hep had already determined that this was not a capital case. It was a "property" case. Hep wanted to remove enough property to really hurt. It was ill gained prosperity.

Suddenly, a smile creased the old face. The irony of the plan that fell into place was a source of the humor. The training he'd gotten in his first military mission decades earlier would serve him in his Final Mission. When he'd arrived in Korea, as a PFC, decades ago, he'd been trained as a truck driver and assigned a fuel truck. He hadn't been particularly happy about that duty. The survival rate wasn't good inside a slow moving, explosive, prime target. However, his good looks saved him. He had been standing guard duty when a general walked by. Hep saluted. The general, stopped and demanded Hep's name and unit. He couldn't figure out what he'd done wrong. The next day, orders came down transferring him into the general's guard. The officer wanted to be guarded by good-looking guys, who had manikin physiques.

Hep lingered over his Korean memories. Actually, that was a hazard-fraught period when he had felt more alive than any other time in his life. There have been death all around, but it was not like the death of his son or his wife. Those events had tainted his life.

Returning to the present, Hep finished his drink and headed for the restaurant to celebrate his mission plan.

In the morning, after a stint on the phone, Hep headed out to look over the bulk fuel depots in the area. He spent the early mornings and evenings watching the various activities.

Hep was particularly interested in the comings and goings of the drivers and the office staff. The plant security was also critical.

Finally, it was time to move. In the afternoon, Hep packed his little bag. He put the nearly spent bottle of scotch on top. He transferred the bag and a sign he had made, into the car before turning in his key and paying his final bill. Although he would've liked to stop at a bar, he didn't.

Instead, he headed across town to a Chinese restaurant. It'd been a long time since he had eaten Chinese. He and his wife used to enjoy an occasional trip to such a place to celebrate special events. Since his wife died, Hep had savored the memories but not the food. This evening he felt especially close to Alice and this was a special day.

Hep had a leisurely meal of both food and memories. He didn't have to be at his destination until after dark.

Fortune smiled on Hep. He had found a bulk petroleum depot at the edge of town on the same side as the Mata ranch. He wouldn't have to drive through town in a stolen gas truck.

The Rinder Petroleum Products stood in an open space. At one time it had been on the edge of town. Then the urban expansion had jumped the fuel tanks, but didn't get too close.

From his earlier surveillance of the establishment, he found out that the truck drivers/delivery men returned to base at various s times in the afternoon. They were out of there by 5 PM. The small office crew left at the same time, leaving two late workers to refill the trucks for the next run. The remaining workers were usually out of there by 9:00 at the latest.

Hep cruised by Rinder's after dark. There were only the two vehicles that were owned by the late crew. He knew both lived in town, so Hep parked where he could watch for vehicles to leave.

At 8:45 both late workers headed into town. Hep adjusted his position and returned to Rinder's. A double, high chain-link gate was secured by a chain and padlock. Hep rolled up to the gate. He left his parking lights on.

Earlier, Hep had purchased a bolt cutter. It was larger than was probably necessary, but he needed the extra leverage to make sure his emaciated frame could complete the job.

Before long, Hep was parking his car in the employees' lot. There were numerous nightlights to illuminate the area. He had preselected a medium-size gasoline truck. He checked the truck number before heading to the office. Inside a handout window was a large wall-hung key box. Beside the window was a wooden door with a glass pane.

Using the bolt cutter, Hep broke a light bulb above the door and then used the rubber grips to knock out the glass window. He was not particularly concerned about the noise of shattering glass. Most of it was contained in the old frame building and the neighbors were distant.

Hep turned on the office lights as they would have been before the late employees left. The lock on the key case wasn't a problem. Before long Hep had the truck fired up. When he left the lot, he draped the chain across the gates.

Hep flipped off the lights as he approached the Matta compound. There was enough moonlight to navigate. After he turned into the drive, he let the tanker drift to a stop on the incline and set the handbrake so no brake lights would come on. Leaving the motor running, Hep slipped out the door. Fortunately, there was no dome light. On the side of the truck was a rack of 4-inch sectioned hoses. Hep tried to lift one down, but he no longer had the strength. He mustered his strength and tried again with the same result. A tendril of fear started to make its way through his guts. Was he no longer able to fulfill his mission? He moved down the rack to the end and was able to wrestle the heavy coupling over the edge. The segment fell to the ground. Hep was able to drag it around so he could hook it to one of the valves below the tank. When he opened the valve, he heard the gushing of gasoline through the hose and when he smelled the stink of the fuel, he breathed a sigh of relief and headed for the cab. With a grim smile on his face, Hep shifted into gear, turned on all the truck lights and laid on the air horn as he began his run up the hill. Before he reached the house more lights started coming on and people opened windows and doors.

Hep headed for the right corner of the main house. He drove close enough to the house that gasoline sloshed up against the foundation. At the back corner, he made a sharp left turn so he wouldn't run into the swimming pool. After dousing the rear of the house, he paralleled the long line of added apartments.

By this time there were adults and kids all over the place. He could hear the cry, "gasolina, gasolina."

Hep drove the truck around the end of the apartments and then down the back side. He jerked a left to take care of a front of the house. Passing on, he made a circle of the garages and then the warehouse.

Having soaked all of his targets, Hep drove a considerable distance out into a vacant field. Before closing the valve, he soaked a washcloth he'd taken from the motel in the gasoline. He moved the truck away from the gasoline soaked ground. With the headlights pointed back toward the house, he put the transmission into neutral and reached for the scotch bottle. There was one large swig left. He rolled the scotch around in his mouth as he watched the skirmish line of males advancing toward him. Some were carrying guns. They seemed reluctant to fire. They probably were afraid of starting a fire.

Hep stuffed a gasoline soaked washcloth into the neck of the scotch bottle. With a cigarette lighter he'd picked up in a convenience store, he held the container outside the cab as he drove up alongside the gasoline trail. Hitting the air horn again, he tossed the flaming washcloth onto the saturated dirt.

As soon as the fire started, all of the males broke back for the houses yelling "fuego, fuego."

Pulling a safe distance from the inferno, Hep picked up real estate sign he'd stolen from a For Sale house. He'd reversed the cardboard insets and on the bare surface he wrote "THIEF." He stuck the sign in the ground.

Returning to the cab, the old soldier pulled a Purple Heart and a Bronze Star from his bag and pinned them onto his shirt. His old hands wouldn't work the keepers. That didn't make any difference. He let them drop into his lap.

The flames had swept around the warehouse and the garage and were headed for the main house. With all of the flame, the tanker was clearly visible. Groups of children were milling around far away from the fire hazard. Adults were darting in and out of the residential structures trying to salvage valuables.

The flames hit the main house. They were engulfing the house from both front and the back. A smile turned up the edges of Hep's mouth as he put the tanker into motion. By the time he reached the swimming pool, he was doing 40. Just before the tanker plowed through the glass sliders into the family room,

Hep uttered his version of a battle cry.

The headers over the glass doors ripped into the top of the tanker creating a heavy explosion and a gigantic fireball.

CHAPTER 10

By the time the diminished roundtable group met, everyone had heard repeated accounts of the after-dark attack on the Matta family in Laredo, Texas. The story fascinated the public because of the bizarre nature of the event. None of the Matta family or their employees had been hurt, but the entire compound had been completely destroyed. The perpetrator had apparently died in a suicide crash, but the fire was still too hot for investigators to look for a body.

There was much speculation over a sign stuck in the backfield. The Mattas said that they had no idea what the sign meant and denied knowing the presumed assailant. It was all a big mystery.

"It won't take long until somebody comes around here inquiring about Hep," said Chucky. I'd suggest we say that he'd been talking about going to visit an old friend before it was too late. We just assumed he'd gotten a call.

"We don't want anyone making any connection between Dane and Hep and Matta," said Merv. "That could complicate matters."

"For that matter, Arn's situation may raise an eyebrow or two," offered Will. "The less we have to say, the better. Ignorance is sublime right now."

"I better have a little chat with Babs. She's already shown an interest in our business. As soon as Hep's story comes out, she'll start putting things together."

"Good thought," said Dane. "Hep has given me a peace of mind I never thought I could get. It's time I return the favor even knowing he's not around to celebrate."

"It might be wise if you hung around until we are questioned about Hep. Another disappearance might raise some more questions," said Chucky.

"I'm in no rush. I want to visit Arn before I go."

"I went to see him yesterday," said Chucky.

"How is he feeling about his trade-off now?" said Will.

"He seems to be settling in. He can handle the daily routine all right. The doctors have taken him off any work list, so all he does all day and night is read, work sudoku or watch TV. He is pleased he doesn't have to pay any bills, buy any drugs or food."

"How does he get along with the other prisoners?"

"He's so old and decrepit that they're afraid to touch him because he might die and they could be charged with his death. Besides, he told them his hands were that way because of leprosy."

Everybody laughed.

Chucky hung around until the others left and then ordered another beer. When Babs brought it over, Chucky leaned forward and although none of the gang at the bar could have heard over TVs, said in a conspiratorial whisper, "Sometime soon there will be people nosing around here asking about Hep. He was the one in that Texas firebombing on TV. As far as we know, he was off visiting an old friend. We're speculating that it was an old army buddy who didn't have much time left. It wouldn't be helpful to volunteer very much information."

"That was Hep?"

"Yep...Hep."

"And what about Arn?"

"Let's not connect to the two, huh."

Chucky could see the conspiratorial image rise in Bab's demeanor and expression. "What's this all about?"

"You don't want to know. You can't be held responsible for information you don't have. Just keep your eyes open. I think you're being missed at the bar."

A couple of days later, a sheriff's deputy showed up while the four remaining Roundtable members were in session. Babs promptly directed him over to the table.

"I'm Deputy Hal Martin. We have an inquiry from Laredo, Texas, concerning a Cornelius Heptner. I went to his apartment and the manager said that the only place he ever went socially was the American Legion."

"Has something happened to him?" said Chucky.

"Yes, there hasn't been any public announcement, but Heptner was the driver of that fuel truck in Laredo that has been on all the news."

All of the Roundtable's pessimists exhibited disbelief.

"Our old, decrepit Hep?" said Merv, expressing amazement. "I took him to the Orlando airport the other day. I was under the impression he was going to see an old buddy somewhere around Cincinnati."

"Did you go into the terminal?"

"No, I just dropped him off in front of the terminal." said Merv.

The deputy inquired if Hep had ever mentioned José Matta or Laredo, Texas. No one could recall anything.

The deputy continued for some time. The roundtable was very helpful but didn't provide much useful information.

After the deputy left, Merv said, "I think I'd better go home and get rid of all that research I did for Hep. I don't really care if it is eventually dug up but I don't want such a thing to interfere with my mission."

"That's a good idea," said Chucky.

CHAPTER 11

A couple of days after the deputy had been there, Hep's name was released. The roundtable became a source of interest to the other legionnaires, but no one was brazen enough to come over. They just asked Babs, who insisted she knew nothing.

Midday attendance grew substantially. There was an overflow from a bar into the tables. Everyone was trying to eavesdrop, but nothing was gleaned since the pessimists took delight in conducting inane conversations.

The next day, Dane failed to show up. This caused additional speculation among legionnaires. The commander, who seldom came in that early in the day, put in an appearance. He wandered over to the table. "Your little group has had a run on bad luck."

"When you get to be our age, there are no guarantees," said Will.

"Where is Dane?"

Chucky shrugged, "maybe he's having a bad breathing day. If he catches a cold, it takes him six weeks to get over it."

"I think I'll drop by his place to see that he's all right."

"Suit yourself," said Chucky. "If he needs some help he usually calls in."

By the time the commander stopped by Dane's apartment, the old man was 200 miles north. He was making good time in his ancient Ford Taurus wagon. He wasn't going to make it all the way to Atlanta in one shot like he would have done in his younger days, but that was all right.

By mid-afternoon, Dane had put in all the time he could afford behind the wheel. He was rather pleased with himself. Only occasionally did he have to turn on the oxygen tank on the seat beside him and slip the cannula in his nose. Before he left home, he had his six shoulder tanks filled and he had rented two larger cylinders that he had to drag behind on a cart.

A strong, young neighbor had helped him to haul his concentrator out to the Taurus. There was just enough room for it to stand up in the back. He had chalked it in with cinderblocks.

One of the reasons he didn't take the freeway system is that he wanted to find one of the older motels where he could park next to his unit.

It took a little searching to find the right place, but eventually he had what he wanted. The place was old enough that it still had garages next to the units. He could run an extension cord and hose out to his concentrator thus having oxygen when he needed it without having to move the unit.

The next day, Dane pushed on into Atlanta. Before he found a place to stay, he looked up the office and home addresses of Dr. Harden James. Hep had followed the doctor's career. Shortly after Hep's complaint, Dr. James had completed his residency in Oregon. His record was clean but his drinking problem had become public knowledge. Hep had speculated that this was the reason that he had moved across the country to Atlanta. He'd married and settled down to stay a while.

The office was in a hospital annex where it would be difficult to accost him alone. That really didn't matter, because Dane was more interested in the wife, if the marriage was still good. This matter needed some investigation. He didn't want to simply rid the doctor of an unsatisfactory spouse.

Besides being a mother of two grade-school-age boys, she owned a floral shop located in a small strip mall.

In the morning, after getting the boys off to school, Evi James drove to her shop, getting there at 8:30am. The business opened at 9:00am.

Evi covered the store from 9:00 to 2:00, when a clerk arrived to carry on until closing. At 2:20pm the boys got out of school. Their mother picked them up and drove them home.

As far as Dane could see, the marriage appeared to be sound, so he proceeded with his plan.

When Evi arrived at the shop, she parked her snazzy little Honda against the back wall of the shop to the right of the rear door. The left side was more spacious. It was reserved for delivery trucks.

Across the alley was the parking lot of a large retail store, which faced onto the next street. One of the aisles of the lot lined up with the back door of the floral shop.

In Atlanta, Dane had checked into a motel so that he could have a bellhop put the concentrator in his room. On the evening of the third day, Dane felt he had enough information. He bought a couple of beers and returned to his room. However, before relaxing with his beer, he made a slow trek to the office to mail an envelope to Dr. James. The operation was engaged.

At 8:30 in the morning, Dane sat in his Taurus in the parking lot behind the floral shop. He was taking oxygen from one of the larger tanks. He didn't want to become lightheaded at the wrong time.

As Evi turned into the alley, Dane started the old station wagon. "Thanks, Hep. This one's for you," said Dane as he dropped the car into drive. At this hour of the morning, there was very little traffic in the parking lot. Dane lined up the target and waited until the Honda came to a stop. If everything was according to routine, Evi would open the door and put a container of coffee on the roof before collecting her purse and tote bag.

She would juggle the bags, coffee and keys to the door and fumble with the lock. She wasn't aware of the oncoming Taurus until an instant before the impact.

The heavy, lumbering wagon embedded itself halfway into the building. A spark ignited the oxygen filled interior of the Taurus. The oxygen tanks added their energy to the conflagration. Soon the fuel tank exploded.

It was a spectacular pyrotechnic display, with obvious news potential, especially with a double fatality. It took a while to find out who the driver had been. The motel clerk called in a tip that an old man, who drove an old Taurus and had oxygen cylinders with him, was registered there. Armed with the identity of the driver and the gleaned description of age and physical condition, authorities speculated that a medical condition might have contributed to the accident.

A big deal was being made that the victim had been the wife of a medical doctor and mother of two young boys.

However, the arrival of a letter mailed to the doctor the day before the accident shed doubt that it was an accident. A nurse opened the envelope and showed the sheet around before the doctor saw it. The envelope contained a single sheet of white letter size paper. Printed across it in a shaky hand was:

A WIFE FOR A WIFE

YOU DRUNKEN BASTARD

Although Dr. James denied any knowledge of the meaning of the letter, authorities began making inquiries.

CHAPTER 12

One of the guys from the bar was holding the legion door open by the time Chucky wheeled up. For the last few days, Chucky had been the recipient of unprecedented attention. He, Merv and Will have been under constant surreptitious observation ever since another member had come up missing. The commander had been the only one to venture over to the Pessimists' Round-table to report that Dane and his car were missing. Babs still hovered.

Apparently, no one had yet connected the dead man at the Atlanta floral shop with Dane. When Merv entered a few minutes later, he raised an eyebrow to Chucky, who returned a miniscule nod. After Merv maneuvered his bench into position and settled his weight, he and Chucky raised their beers in a silent salute for a successful mission.

Babs delivered Will's drink before he made it in the door. She put it in his usual position, but Will picked up the glass and moved into Dane's former location to Chucky's right. Merv was now on Chucky's left.

Actually, Will's move was prompted by the desire to talk privately, but Babs took it as a sign that Dane wasn't coming back. She turned her peripheral attention to the numerous TV sets scattered around the room. She monitored the newscast. When the report of a fiery death of a floral shop owner and an elderly driver came on, Babs immediately whipped around and mouthed, "Dane?"

Chucky held his beer can up for a refill. When Babs arrived, Chucky said, "Yes," but you don't want to know. There will probably be people nosing around. Just pass them off to me."

When Babs had left, Will said, "How soon will they be snooping around?"

"Since they know it wasn't an impaired man, they probably know who he is and where he's from. It won't take long," said Chucky.

"In that case, I'd better pull out before anyone gets here. Merv, do you have any more info that I should know?"

"No, you got it all. How are you fixed for money?"

"I'm fine. I can't see any reason for paying another month's rent. I won't need much."

The next day, Chucky found the legion parking lot filled with cars. The membership had apparently found out about Dane. Everyone would know that something was up. The pessimists' table was still vacant, although most of the other tables were occupied by the overflow from the bar.

When Chucky arrived, there was a noticeable reduction in the noise level. The reaction of the membership amazed Chucky. Here were a bunch of guys, many of whom had been in battle and they were afraid to confront the roundtable with their questions. When it became obvious that Will was either very late or not coming, the buzz got louder. Finally, one of the regulars at the bar yelled, "Where's Will?"

"He said he was going to Audie Murphy to see if they could give him something to make him young again," shouted Chucky across the room. The comment produced a hearty laugh from all those who knew Will.

An elderly, overweight man with a game leg detached himself from one of the bar groups and hobbled to the roundtable, relying heavily on a cane.

Without an invitation, the man fell into a chair across from Chucky and Merv. "My name is Bert Bates. I'm from post 180, down the road. The word that's wandering around down there is that you're having an unexplained loss of membership."

"Oh I don't think it's unexpected," said Chucky. "They were all old man and not in the best of health."

"Yeah, sure. None of them died in bed."

"That would be a hell of a way to die," said Merv.

Since Chucky hadn't sent Bert packing, a couple of older local members sidled up to the table. "Where did Dane say he was going?"

"He didn't say," said Chucky. "I figured he was at the doctor or having a breathing problem. Why don't you sit down and take the load off your feet?"

The three guests continued to probe concerning the strange fates of the departed legionnaires. Finally, Merv said that he had to get some things done. As he removed his bench from the table, Chucky said, "Yeah. I have to go too. Will you hold the door for me?"

"Sure."

Chucky followed in Merv's wake. "I was going to tell you in there, except we were interrupted, that I am about to spring my mission, but I can do it from here," said Chucky. My target is in Des Moines. I can do what I need to do over the Internet so there is no reason to travel the distance. Also I have set it up so that no one will connect me with it. I think we may have started something going here. I'll hang around a little longer to see if it can be developed into a viable program."

"I have my mission lined up too. I'm not quite sure I can sell my story, but it doesn't really matter. I'll wait around until I see how Will fared. The authorities will start connecting the dots as soon as Will is identified and then they'll be looking for you and me."

CHAPTER 13

Will took a last look around his skimpy digs. There was a satisfied smile tugging at the corners of the slit that was his mouth. The door latch clicked but he still gave the door an extra kick with the heel of a shoe as he turned away.

The drive to Jacksonville was uneventful, though tiring. Will was glad when he finally fell into an Amtrak seat. The trip to New Orleans would only take a day. He intended to enjoy it as much as possible. Arn had taken care of Will's most bothersome, unresolved problem. He was delighted at his calm mental state as he started his Final Mission.

As the world clicked by, Will mentally reviewed the target's history. After Merv had raised a stink about bad drugs being dispensed by the crooked pharmacist, more information surfaced closing the pharmacy. Van Arden came out of the whole affair without his medical license being revoked, but his reputation had been severely tarnished.

The community had too long a memory for Van Arden to start over again in that part of the country, so he moved to New Orleans where he concocted and then marketed the world's most amazing joint pain remedy. According to what Will could find out, Van Arden was using his medical status to legitimize the product. He was also a glib promoter. Reports added that the business was brisk and expanding.

Will was convinced that this was just another quack with another brand of snake oil.

After a little doing, Will settled down in an older motel with a chain restaurant on the corner. Across the street was the Van Arden Famous Joint Remedy home office and plant. From either his room or the front booth of the restaurant he could watch the activities at Van Arden's.

In time, Will identified the target and the various office workers. The factory workers, who numbered a dozen or less, parked in the back of the building. The office staff cars were in a small parking lot in front.

Van Arden was always a late arrival. A couple of men who were usually dressed in suits, came and went at irregular intervals. Will figured that they were the marketing staff.

Van Arden usually worked until 7:00pm or so and then met someone for a leisurely dinner. From his dress and wheels, Will figured the doctor had found another profitable enterprise.

Will had been watching the comings and goings all afternoon. It was Thursday. Will figured he'd make his move on Friday, just before quitting time. Workers were usually anxious to leave for the weekend. Maybe some would have already left. For his purpose that really didn't make much difference.

As Will adjusted his position in the chair to ease his pain, he watched the two guys in suits leave. Then one of the women left, while putting a bank pouch in her bag.

Will abruptly stood up, pulled an old wooden box from the dresser. The box looked like a long cigar box. Will tucked it under his arm in a horizontal position and headed for the door. Aloud he said, "Why not now?"

At a measured pace, Will jaywalked across the street and straight in the front door. A middle-aged woman, seated behind a computer, met Will with a minimal smile. "How may I help you?"

"I'd like to see Dr. Van Eden for a few minutes. It won't take long for him to decide if he's not interested or if he wants me to hang around."

"What reason can I give?"

"Tell him it's about a Mexican herbal."

The woman started to pick up the phone, but thought better of it. Instead, she walked down a short hall to the end office where she tapped on the door while keeping an eye on Will. A short time later, she came back. Dr. Van Arden will see you. Go right in. He's expecting you."

Dr. Van Arden was standing behind a large executive desk. As Will entered the room, he said, "Good afternoon Dr. Van Arden. My name is Will Longknife."

"Mr. Longknife. Please sit down" He indicated a comfortable looking chair in front of the desk. "How may I help you?"

Will seated himself and balanced the wooden box across his bony knees.

"When I was younger, I used to travel rather widely in Mexico. I became interested in native remedies. There are a bunch of them. For instance there is a cactus root, which is soaked in Mexcal. A little of that extract will deaden the inside of your mouth almost instantly. I've used it on toothaches.

"You have something you want to show me."

"Oh yes," said Will, as he stood so he could place the wooden box on the front edge of Van Arden's desk, he positioned it so that the raised lid shielded the contents. Will reached in with both hands. Dr. Van Arden sat up in his chair expectantly.

Will adjusted his grip and as he brought his hands up, he cocked a brace of .22 Ruger Bearcats revolvers. Before the doctor had a chance to react, two shots sounded. A pair of little lead slugs entered both the doctor's shoulder joints. Simultaneously, clicks of the hammers on the single action weapons joined the anguished groan from the other side of the desk. The guns barked again. This time the targets were the elbow joints. The second volley brought a cry of distress and panic.

The impacts activated the doctor's flight mechanism, causing the chair to crash into a cadenza against the wall. Both the doctor's arms hung at his sides giving Will clear shots at the hands. The slugs went in, cutting across the back of the hands, messing up all those little bones.

A frightened cry came from the other side of the door, "Dr. Van Arden, Dr. Van Arden."

Dr. Van Arden's face revealed his pain and anguish. He was still conscious. Will cocked the Bearcats then leaned forward, saying in slow precise words, "This is for the all the pain and anguish you've caused with your overprescribing of bad drugs... you quack." With that, Will blew out both kneecaps. He hesitated for a moment to watch Dr. Reynard Van Arden passed out.

In the distance, there was a siren. Will had stepped around the desk to get shots nine and 10 into the doctor's ankles. Without even a split-second hesitation, William Howard Longknife cocked the pistols and jammed both barrels into his own eyes and simultaneously pulled the triggers. No arthritic pain jabbed him when he hit the floor.

CHAPTER   14

As soon as the reports of the bizarre shooting and suicide hit the news, Merv pulled on his shoes and picked up his car keys from the bowl on the dining bar. Someone was bound to start connecting the dots and he didn't want to be caught at home sleeping. A smile played on his lips as the radio announcer describe how doctors were frantically trying to repair the massive damage created by the .22 caliber long rifle bullets when placed in articulated places.

As the unbalanced car moved through the darkness of South Florida, Merv was especially savoring a story that Will had told a long time ago. Will and his wife spent a lot of time in the western deserts. They enjoyed the outdoors and the rugged beauty of the arid areas. Also they treasured the uninterrupted time together.

Being miles from any humanity or domesticated animals, Will bought a Ruger .22 Bearcat. It was a small, single action revolver...a frontier model. Will loved to plink at things. His wife, Alice, found she enjoyed plinking as well. To salvage his own enjoyment, he had to buy Alice a duplicate.

They became excellent shots. A fun game for them was to march a tin can across the countryside shooting at it alternatively.

Will had put these skills to good use. Dr. Van Arden would quickly learn what it is like to live with pain and disability. Will had done a worthy job. Van Arden's disregard for the well-being of others had consequences. Merv savored the thought.

As Merv pulled into the outskirts of Venice, Florida, his thoughts were dragged back to the mission at hand. He'd made maps from the Internet. Wesley Holmes lived curiously close to his place of employment.

Since it was a workday, there was a good chance Merv would be able to spot his target leaving for work.

Holmes's residence turned out to be a small two-story, four-plex on a corner. Across the street was a Denny's restaurant and on the other side of the restaurant was a strip mall that contained a number of commercial spaces, including Ace Accounting, Holmes's employer.

Merv figured Holmes would walk to work, so he pulled into a public parking lot across the street where he could see both addresses. He eliminated both ground-floor units. Lights went off or curtains closed before the occupants left. They were women.

If Holmes lived on the second floor, that would eliminate the apartment as a point of action. Merv didn't want to attempt that many stairs.

At 9:15 AM a young man in a suit exited the building. He matched the vague description Chucky had supplied. The man walked across the street to Denny's. He was inside until 9:50 AM. Then the man moved across the parking lot to Ace Accounting.

Merv gave up his surveillance in favor of finding a motel, which was always a problem for anyone of his weight. He had to find a manager who would remove the bed frame and put mattress and springs in a corner. It was difficult getting on his feet from such a low platform, but it was better than disentangling himself from a broken bed. At home, he had two courses of concrete blocks on the floor to reduce the amount of energy needed to get into a standing position.

After a nap, Merv went through a fast food drive-through instead of wrestling with seating in a restaurant. He returned to the parking lot in front of the target's workplace.

Ace Accounting was a busy place with lots of people coming and going. He could see that there were numerous small cubicles off a central corridor. Merv's initial estimate was that there was too much traffic and the area was too open to serve his purpose.

A little after 5:00 pm, Holmes left the office to return home. Merv shifted over to the lot in front of the apartment. He'd wait to see if there was a dinner trip.

Dressed in casual clothes, Holmes turned left when he hit the street. He ended up in a small family style diner a couple blocks away. After dinner Holmes stop by a neighborhood bar. He didn't leave until 10:00pm. He was pretty unsteady by the time he made it home.

Merv headed back to his motel. He was pressing his luck by hanging around Holmes's place so much. His left-tilting car was too conspicuous. He would have liked to spend a few days watching his target but he needed to act soon. In the morning, Merv was in Denny's early. He found a bench along a wall that would handle his weight. On the way in, he looked over the glass brick entryway. Both the inner and outer door swung out. It wouldn't be ideal, but it could work.

Merv made a trip to the men's room. It was much more suited to what he had in mind. The door swung in to the left. There was a short hall. To the right was a utility closet and straight ahead was the stall. To the right of the toilet enclosure was a urinal on the back wall, and against the right wall was a wash basin.

Holmes arrived precisely at the same time as the day before. He sat on the far end of the counter where he joshed with the waitresses. Merv was separated from the counter by a partition, which was low enough for Merv to see over. Holmes finished his meal and called for a refill on the coffee. He then headed for the restroom just like a cat heading for the sandbox after eating.

Merv had been watching the toilet traffic. It was rather sparse. No one was in there when Holmes headed in that direction. Merv started moving his bulk. When he reached the door he couldn't hear anything. Opening the door, he could see shoes under the stanchion. Merv backed out and stood as if waiting his turn.

The toilet flushed. Merv waited a few more seconds before shoving the door open. Holmes was pulling a towel from the dispenser. Merv stood behind him as if waiting to get by to the urinal. Holmes glanced at who had come in. A little shadow of contempt passed over his face. As he turned to head for the door Merv thrust his weight forward crushing Holmes against the closet wall. There was a great whoosh of air as all of the wind was knocked out of Holmes.

Merv grabbed the breathless accountant by the lapels throwing him onto the floor on his back. Then Merv dropped his full weight across the upper torso. Holmes hadn't been able to utter a sound. Merv maneuvered around so Holmes's face was close to his.

Holmes eyes were bugging. He was trying to suck in air but the several hundred pounds kept his lungs collapsed.

Before Holmes lost his senses, Merv snarled, "Cowardice in combat has its consequences. This little hug is from Sergeant Chucky Bentley."

Holmes's eyes widened even more when he realized this was payback time. He had a few more seconds to contemplate the situation before passing out.

Merv adjusted his position so that his feet would permit the door to open only about 4 inches. Nothing could save Wesley Holmes now. Even if Merv wanted to, he'd be unable to shift enough weight off of his target before the coward was long dead. When a hand hit the restroom door, Merv relaxed into a feigned unconsciousness. A guy at the door tried three times to bang the obstruction out of the way. Finally, he put his shoulder to the door and shoved. His effort was not enough to push several hundred pounds across a non-slip tile floor.

From what Merv could piece together the first guy called the manager, who tried the door and then went for a hand mirror.

"Hey. There is a mountain of a man lying on the floor. Can't tell if he's alive or dead," said someone. Several more looked through with the mirror. Then someone stuck a phone camera over the door. That was when they discovered that there were two bodies on the floor.

After some discussion, a lawn care guy brought in a small chainsaw. He cut across the door about a quarter of the way up. Sirens were approaching rapidly. Merv decided to start regaining consciousness before paramedics arrived.

Someone was trying to find a pulse on Holmes. They couldn't get to his hands and his neck and half of his head was enveloped in fat.

Merv gave a little groan and twitched a tiny bit.

Someone observed, "I think he's coming out of it." The observer repeated his observation to the rescue team as it arrived.

Gradually, Merv appeared to regain sufficient control to expose enough of Holmes so a conclusive determination could be made that he was indeed dead.

Ultimately, Merv told everyone to stand back so that he could maneuver in that confined space. With difficulty, Merv moved into the open area in front of the stanchion. He made it into a sitting position with his back against the wall.

CHAPTER 15

The day after Merv left town, Chucky didn't bother to go to the legion. He needed the midday hours to start his telephone campaign. Earlier, he had purchased a pre-paid cell phone. He never put any correct personal information into the system. Using the new cell, Chucky called retired detective, Lieutenant Arthur Bidweiller in Lake Oswego, Oregon.

When the retired policeman came on the line, Chucky said, "Lieutenant Bidweiller, I understand that you spent a good many years looking for 'The Trucker'."

"That bastard," snarled Bidweiller. Who are you and what do you have to do with that SOB?"

"I'm not important, but for convenience call me Jason. Do you still have enough fire backed up to reel him in if I give you a name and address?"

"You bet." Chucky recognized an instantaneous change in the voice with only two words. If Bidweiller didn't have the energy at the moment, he would generate it. He caught fire, just as the guys at the Pessimists' Round Table had at the thought of a Final Mission, for a worthy cause.

"The Trucker' is living under the assumed name, Arnold Leonard Leitner at 820 Pinecone Lane, Des Moines, IA. I understand he's a nasty piece of work. Do you still have enough contacts to nail him?"

"I still have the contacts. How can I get in touch with you?"

"There is no need to get in touch with me. You have all the pertinent information that I have. I may call in occasionally for an update to satisfy my curiosity. Happy hunting." Chucky pushed the button.

Chucky figured that one call was enough. If Bidweiller had Alzheimer's, a stroke or something like that, he'd have had to make multiple calls and probably more personal involvement. He decided to go down to the legion after all.

The parking lot was full. When Chucky wheeled himself along the ramp, someone he didn't know was on hand to open the door for him. As he passed through to the inner door into the bar he was greeted with a chorus of "Hi, Chucky." All of the bar stools were occupied, as were the tables. The one exception was the roundtable. Everyone watched as Chucky wheeled to the table and swung into place so he could face the crowd.

The commander, Burt Stark, detached himself from a group standing by the kitchen door and came to the table. Babs hustled around the end of the bar with Chucky's beer so that she could be within earshot of what the commander had to say.

"Hi, Chucky. Where's Merv?" said the commander as he sat down to Chucky's right so he wouldn't have to shout across the table.

"He's been talking about getting some medical attention. He might have gone looking into that."

"There was a deputy sheriff around here this morning wanting to talk about Will. Babs said he'd been here before. We're beginning to get an odd reputation."

Chucky snorted. "It looks as if an odd reputation is good for business. There hasn't been this many people in here since the end of World War II." Chucky scanned the congregation and Chucky and the commander were the only subjects of interest.

The commander didn't care for the big audience, but he didn't want to appear secretive and take Chucky outside. They couldn't go to his office, because it was upstairs. "The deputy said he'd be back. He was going to go have a look at Will's apartment."

"I don't how I can help him, but send him over to see me if you think it might do any good."

"This little post has Arn being held for murder and two, maybe three, spectacular suicides in a month's time. Those figures defy the odds and as soon as the police connect the dots, they're going to figure our post is the headquarters for Murder Inc." Stark dropped his voice as he put a beer can in front of his face. "I want to know what's going on."

Chuck duplicated the action and from behind his can, he said, "No you don't."

"People are starting to forward to me a whole bunch of stuff from the Internet. The conspiracy theorists are beginning to tune up. They will have a real field day with this."

"That's all right. Let them go off on any flight of fancy they want. It'll be good for business.

The conversation was interrupted by Babs coming over to hand her boss a list. "Better get this stuff over here as soon as possible. We're almost out of everything. You'd better triple the food and get Hal over here on the grill. I can't handle the bar and the tables plus cook hamburgers."

"I'll get right on it," said Stark. To Chucky, he said, "We'll talk later."

As soon as the commander moved away, Bates, the old man from the post "down the way" lurched over to the table. Hi, Chucky. Remember me?"

"Yeah, sure Bert. What's up?"

"You've lost another member since I was here. At this rate you'll have to close the post soon."

"Either that or expand to it to handle all the other old duffers in the region."

"Say, I've met a couple of those from different posts. They're at the bar. They'd like to meet you. Mind if I invite them over?"

"Seems as if I have an abundance of chairs today. Go ahead."

Bates raised his bony hand and flicked a finger. Two old guys slid off barstools, picked up their beers and slowly moved in the direction of the table.

The lead guy sported a handsome shock of white hair ringing an ancient face that was set with what looked like a painful grimace. He leaned heavily on a stout cane. His right leg was badly misshapen.

"This is Dale, from post 189," said, Bates as the body with the head full of white hair settled into a chair. "This is Karl from post 172, up the beach. Both of you know who Chucky is. The three of us came to the conclusion that something interesting was happening down here. We're tired of the game we're playing and were looking for a new one.

Chucky looked from one to the other of the occupants of the table. He felt they were all sincere, but it didn't make any difference. He couldn't help them.

"I'm sorry, guys. I have no new game. The boys and I are always discussing various diversions but everything we've come up with has developed major flaws. That's how it stands."

Karl looked disappointed. "Maybe we can help you 'deflaw' your program."

"No. All this interest," said Chucky, as he swept his hand toward the bar, "creates problems, which will have to be worked out. In fact, you'll have to excuse me. One of those problems is just coming in the door."

Deputy Baca surveyed the scene. The commander wasn't inside. Baca did a rough count of the number of people squashed into the bar. After his calculations were complete, his eyes slung back to the Pessimists' Roundtable. As he bore down on Chucky, the three guests at the table figured it was time to go.

"You're Chucky Bentley, right?"

"You've got it. Sit down. I don't imagine you'd have a beer, so I'll offer you a soda or a coffee...if you think you can handle it. By this time of day it is snarling loudly enough that Babs has to put it in a cage in the kitchen."

Baca smiled. "Suppose we might make some side money by putting legion coffee up against sheriff's office coffee?"

Chucky laughed. "From what I've heard, we wouldn't have a chance if we had to go up against Second Night's brew. What can I do for you?"

"You can tell me what's going on around here. I've been sitting out in the parking lot for some time taking down information concerning a fatality down in Venice. It happened earlier this morning in a restaurant where a man was squished to death by the guy that normally sits on that big sturdy bench."

"Merv Dawson? Are you sure?"

"Well, there aren't many 600-700 pound men walking around carrying Mervin Dawson ID and a membership card for this place. I think we're pretty sure of the identification. What was he doing down there?"

"I have no idea other than he found some place down around Naples that he was looking into for weight loss help."

"The reason I'm here in the first place, is that we're being asked to inquire into another of your members, William Longknife. He really messed up a doctor in New Orleans before turning his guns on himself. He was supposedly going to Texas for medical attention. This is my second trip out here. I had to do the same thing on another of your members. When I get back to the office and start searching the files and asking around, how many more legion members will I find who have become involved in violent deaths in recent times?"

"Our membership has taken quite a hit in recent weeks. One is in jail. Three others have died and Merv is apparently in South Florida. This isn't a very big post."

"The heck you say," said Baca as he swung around to look at the dozens of faces watching the proceedings at the pessimists' table. Deputy Baca dropped his voice as he proceeded. "You know that each of these jurisdictions that have asked us for assistance are scattered far and wide and they don't know about any situations other than their own. As far as I know no other deputies are curious. Of course some office staffer might handle some of the reports and become interested."

"Why are you telling me this?"

"With what little I know now, I strongly suspect there are forces on the move that would come to light with a little poking around. I have no idea whether or not these are for good or evil. I'm trying to decide on what to do. Any investigation I conducted will find its way into reports. I would be delighted to uncover any dastardly conspiracy but I would also hate to screw up a worthy undertaking."

"Any probing around might be interesting from a sociological or maybe a psychological standpoint but there would be little of interest from a law enforcement point of view. It would seem that any legal concerns would be satisfied with the surrender or the death of the perpetrators."

"There is more to it than that. Law enforcement is also interested in why. When a small American Legion Post suddenly has five of its members involved in major felony events, there is much more to it than just random events.

"You're not planning any out-of-town trips are you?"

"Chucky laughed. "No. I don't travel well." He slapped the arms of his wheelchair for emphasis. Baca started to rise.

"Do you have any idea what is happening to Merv?" said Chucky.

"No. They didn't know if any crime had been perpetrated... yet."

CHAPTER 16

For three days, Chucky fielded questions from curious members and visiting information-seekers. The general theme involved, why had all of those guys done what they did, knowing they wouldn't survive. Most viewed it as a suicide pact but why in this form?

Most of the guys were too young or in good enough health to understand. For them, each day was not a penance that must be endured. Chucky had never heard of a cure for old age.

There was a false assumption that Chucky did not challenge. The thought was that each lost member was settling some old personal debt before it was too late. The possible exception to that was Arn and the drug dealer. His action was purely a business deal.

It was the older men in the post who are getting closer to the truth. Their tomorrows had lost much of their appeal.

Chucky downed the last splash of beer. He wasn't planning on staying for Sloppy Joes. He'd made up enough evasive answers for one day.

"Hey, here's Merv," shouted someone. There was a general movement toward the door.

Sure enough, a few minutes later Merv was giving a good impression of an icebreaker as he plowed through the crowd. He made his way to the bar and held out his hand. Babs filled it with a can of his beer. Merv drained the can and reached for another one.

"You guys have probably heard that I had an accident and that I squished a poor citizen. I'm sure sorry about that. The police wouldn't let me leave until they were satisfied that it was an accident.

That comment brought a few snorts from around the bar. Merv gave a brief account of the incident before he escaped to the roundtable.

Both Chucky and Merv nodded to one another but held off on any conversation until Merv had wrestled his bench into position and arranged his bulk on it.

"Message sent and received," said Merv.

"Thank you. That closes a lot of accounts."

"I know how you feel. Will did the same for me and better than I could've done."

"How come the police turned you loose so soon?"

Merv chuckled. "When I 'came to' and was able to move off the feller, I was laying in the restroom next to the toilet stall. Later, after they had removed the body, they told me to get up. It's not easy for me to get up off the floor. I pulled the stanchion out of the floor levering myself upright. When it went, I lurched against the wash basin and knocked it off the wall, flooding the room.

"They had to take me to the dining room so I could sit on the bench where I had breakfast. Nothing else in the place would hold me."

"What excuse did you have for being in town?"

"I told them I'd been heading for Naples. There is a woman down there that has invented around a dozen exercise machines for people with injuries or disabilities, who can't exercise on their own. She has her own factory to produce her product. Those machines exercise all different parts of the body. She can't tout any of the medical benefits of the machines without having the FDA on her back, so she just builds exercise machines. I have a whole folder of downloads and testimonials.

"I told the cops that I was going down there to look over this place to see if it could help me. I began to feel weird so I pulled into a motel. I'd come out for breakfast. When I went to the rest-room, it was occupied. I had been backed up against the wall to let the guy by and that's all I remember until everybody was standing around."

"They believed you?"

"I don't know if they really believed me or they thought it was to their advantage to believe me."

Chucky raised his eyebrows.

"The medics were going to transport me to the hospital but the gurney was too small and fragile. I'd have had to ride on the floor of the ambulance and they had already seen me trying to get to my feet and didn't want me in their ambulance.

"The medics checked me over as well as they could with me sitting in the restaurant. They pronounced me fit. Of course, they didn't have the facilities at the ER to handle me.

"So they were going to take me to the station. To transport someone in a squad car, all males have to be cuffed." Merv held out his wrists.

Chucky laughed.

"They couldn't even use those plastic straps because my fore-arms taper down the wrists. They would slip off unless they were cinched down tight enough to cause injury.

"Besides they were uncertain as to whether or not I could get into a patrol car, let alone get out.

"I suggested that I could drive my own car. They immediately rejected that idea because I just passed out and killed a man. Eventually, an officer was going to drive me in my car. But when we finally made it to the car, the driving compartment was too big. He couldn't reach the pedals. I ended up driving and the cop was a very jumpy navigator."

Chucky laughed at the visualization. Merv was a very intimidating figure and the cop was going to be a passenger.

"When we made it to the police station, there wasn't any place for me to sit other than on a steel bench in the front waiting area. There were too many people around to conduct any police business. They finally stacked three spare tires and part of a traffic barrier in an office so I could sit down.

"It became apparent the police didn't know what to do with me. They didn't have any jail clothes big enough. Their bunks wouldn't hold me, and they were afraid I'd eat them out of their budget allotment.

"I gave them the keys to my motel so they could get my files on the exercise place I'd told them about. I also asked them to bring back my pillbox.

"They were reluctant to let me take any medication. I showed them a list I carry my billfold enumerating all of my meds, their purposes and their costs. I pointed out what might happen if I didn't stay on schedule. I also told them that I only had two more days' supply and if they hung onto me, they'd have to fill my prescriptions. One look at the prices of all those pills scared the hell out of the Chief."

"I can see they had a big problem. How did they solve it?"

"Around quitting time for the day crew, the chief decided that no determination of a crime could yet be made and they weren't going to hold me, but I shouldn't leave town. That way I slept in the motel, fed myself and worried about my own meds."

"Since you're here, I suspect that they decided that it was just an unfortunate accident."

"Yep. That was by far the easiest and cheapest way out, especially since Mr. Holmes didn't have much of a family and none of them seemed to be particularly concerned."

Merv looked at the crowd at the bar and tables. "What's going on here?"

"It would appear that we have started something. But this is not the place to talk about it. The guys are all waiting to hear your story. Give me a call when you have the time to talk. I can't have you over to my apartment. There would be nowhere for you to sit."

"Come over to my place. There's plenty of room to navigate with that little chair."

"Okay, I'll bring the beer."

CHAPTER 17

When Merv called later, he was concerned. "Half of the guys at the legion were from other posts. What's going on? They were all loaded with questions that I didn't want to answer."

"Are you at home now?"

"Yes."

"I'll come over and we can talk."

"The place is still a mess."

"So is mine and I haven't gone anywhere. I have a six pack in the fridge."

After the two men were settled with their beers at Merv's, Chucky started the conversation. "Our members must have noticed a change in our demeanor after we decided to take on the Final Mission. Remember, they named our table 'the Pessimists' Table.' I wasn't aware of any change, but apparently we were too transparent. We weren't shocked when Arn blasted that kid. We were too pleased...especially Will. Then, one after another of the pessimists went down and the remaining ones were too invigorated.

"It never occurred to me that our little project would create so much interest. Oh, I figured that the locals would figure out that we were up to something but since there would be no connection between the legion in any of the recipients of our attention, it would quickly blow over. If none of us survived, it wouldn't make any difference. As it so happens, you and I are the only ones in jeopardy."

"How do you figure?"

"Since all these initial injustices took place all over the world, no one could connect the victims. And since there was no history of connections between our guys and the ones who received our attention."

Merv snorted, "There are no useful links, especially by the authorities, where the actions took place."

"However, we have one deputy who will probably piece things together."

"Who's that?"

"Deputy Bo Baca. He is the one I had to talk to the day Venice PD inquired about you. All he has to do is Google Holme's name and mine is sure to pop up because of the legal maneuvering after I got back home. That will open the gate and he'll start trying to cross match all of us."

"If you suddenly leave, that would really ring his bell."

"It doesn't seem that I am going to have to go anywhere. I found a retired detective who devoted a good portion of his career chasing 'The Trucker' and never did identify him. I awakened the tiger in the detective and I think he can accomplish what I wanted, many times better than I could. In a day or two I'll call for a progress report. If he can't get him, then I'll have to go to De Moines myself."

"What are we going to do with that deputy?"

"He suspects some sort of collusion or conspiracy, but he's acting rather strangely for a cop. I think it might be interesting to see what he has in mind."

"With all those other guys hanging around, anyone would get suspicious that something is going on. As far as I know, none of us confided in anyone else, so no one should have known about it."

Chucky shrugged.

"You and I are younger than Arn and the rest, so we're not looked upon as being at the end of our rope. There are a few other old guys around here, but most still have wives or family and health problems aren't weighing heavily on them. They are not yet candidates for a Final Mission.

It looks as if the other posts around here have their own version of the Pessimists' Roundtable. As for me," said Merv, "I don't really care what the police do. If they decide to throw me in jail, I'll just stop taking some of my meds and it won't take too long." Merv shifted his bulk to ease some pain. "If they connect the two of us through Wesley Holmes, what can they do?"

"There is probably some sort of conspiracy law by which they could charge both of us with murder. Don't worry about it as far as I'm concerned. I'm so pleased you got the word to that son of a bitch before he died that I don't care what happens. If it looks as if the cops have a case and are going to arrest me, I'm ready. You might say I'm wired."

"How do you think we should handle all these questions?" said Merv, as he tried to find a more comfortable position.

"Until we find out what that deputy is going to do or whether somebody else tumbles to it, let's be evasive. Everything is just coincidence. I think we have a good idea here. I'd like to pass it on."

Merv nodded his agreement. "I think so too, but there are some flaws in the plan that could pass harm onto others. You know the big philanthropist that calls a press conference when he gives a check to his favorite charity is thumping himself on the back instead of primarily being concerned with the recipients of the charity. I more honor the unknown guy who dropped a gold coin into the Christmas charity kettle. I think the 'Final Mission' is going to have to be a completely solo event.

The next morning, Chucky pulled his file on "The Trucker"

and placed a call to Bidweiller. Even though it was early in Oregon, the lieutenant was already up.

"Hi Lieutenant, this is Jason. Did you find 'The Trucker'?"

"You bet we did, just where you said," laughed Bidweiller. He was obviously delighted.

"How did you handle the situation?"

"Back when I was chasing this guy, I made friends with a lot of guys in other departments all over the country. Most of them are now retired, like me. However, we remained connected with the old office. One of the guys was in a little town of Blairsville, Pennsylvania. It seemed as if 'The Trucker' had some affinity for that area and periodically took some poor schmoo for a bundle. Joe Blaine worked the case for years, just like me.

"Anyway, Joe jumped on the tip and drove to Des Moines with all of his files on 'The Trucker'. Armed with all that information he was able to get all of the local law enforcement assistance he needed. They picked up "The Trucker's' fingerprints and identified him as Thomas Purdy, a smalltime con-man with a long rap sheet of minor traffic infractions, battery and assorted con games.

"We have numerous warrants coming down the pike. In a few days, we will be ready to move."

"Good, it sounds as if you have the whole situation in hand."

"All except how you got this information."

"A short time before he died, Arn Leitner told me a story about being on vacation back when he could get around. He was surf fishing with a guy who said he knew a guy by the same name up in Des Moines. Arn found out what he could because years ago he had all of his identification stolen and the fallout from that plagued him all the rest of his life. By the time he received this information, he was too old to do anything about it.

"Before he died, he told me about the whole affair. The other day I Googled 'The Trucker' and boy did I get a lot of info, including your name. Thanks for the report. I'll check back later." Before Bidweiller could object, Chucky broke the connection.

CHAPTER 18

The next morning, Merv was smiling broadly as Chucky was telling what he had learned from Bidweiller. The story was interrupted by the arrival of Deputy Baca.

The deputy seated himself at the roundtable without invitation. "You must be Mervyn Dawson. Venice PD said they had a big problem. I didn't realize how big. I heard they turned you loose. They obviously didn't do their homework."

Chucky pulled himself forward so he can rest his elbows on the table. He tented his fingers and looked over his hands at the deputy. "You're trying to say that you know more than the Venice PD?"

"I can't really blame them because they aren't privy to the raw data that I have, so they couldn't cross reference an assortment of names."

"And?" said Chucky.

"The easiest was you and Wesley Holmes because of the notoriety of the claims and counterclaims between the two of you. It would appear that Merv was your avenger." Baca looked at Merv. "The chances of stopping in a small town because you're feeling ill and falling on the guy who your friend and legion brother thinks caused those gross injuries is rather remote."

"Do you expect to get any admission confirming your speculations?" said Chucky.

"No. I don't expect that. If I build a case, it will stand on its own without having to rely on confessions. At the moment, I'm just trying to scratch a private itch." Baca stood up. "I'd suggest you keep a low profile for a bit. See ya."

Everyone in the place watched the deputy until his unmarked car disappeared from view. Then they returned to their drinks, which was an excuse to watch what was happening at the Pessimists' Roundtable.

Commander Stark came out of the kitchen carrying a fat file folder. He dropped it on the table. "You guys should be getting these things instead of me. I don't have the time to sort post business from the pile of crap."

Stark was obviously miffed as he stomped away. He didn't leave the folder near enough for either man to reach it, so Merv had to move his bulk onto his feet and walk around the table.

As Merv settled down, Chucky began to shift through the sheets of paper. Stark had printed those e-mails that pertain to the various activities of the roundtable.

"What have we there?" said Merv.

"It appears that these are only three days' worth. From what I can see, they're from all over the country and most are from individuals, not posts. However, read some of these and see what they're about." Chucky shoved half of the stack over to Merv.

Chucky motioned to Babs for a couple more beers. Half an hour later, Merv shoved his stack back to the folder. "Could you make any sense out of all this?"

Chucky leaned back in his wheelchair. "No one knows what has been going on here so they're all building images in their own minds. Apparently, there has been some heavy internet traffic concerning our activities. I hope this doesn't come to the attention of the Venice police or other involved agencies."

"How the hell did all of this get started?"

"I think Arn started a lot all of this. There was heavy national news coverage of the dingy old man, who traded his freedom for public eldercare. It seems as if there is discontent over the end-of-life options open to our elders. People don't want to wait around for death."

"Plain old suicide isn't for a lot of folks," said Merv.

"But, guys will sign up for missions where the possibilities of survival are slim to none."

"Those guys figure the undertaking is worth all of the risks..."

"This is different" said Chucky. "Those guys that sign onto possible suicide missions are not old, worn-out men. I would imagine the old or infirm would be the first in line if they could physically handle the mission. They don't have to tabulate all of the good things they would be missing. The old guys would be more concerned with pain and worry relief. To some, no future seems preferable to a future."

From the tenor of the room something had changed. Chucky and Merv turned from their conversation to see what had caused the difference. Following the gazes, Chucky spotted Deputy Baca getting out of his unmarked squad car.

"Hi again," said Baca as he plopped down across the table. "Relax, I'm not investigating anything at the moment. I hadn't gotten out of the area, when I realized I'd forgotten something, so I came back. I was going to ask you if you come down here on Saturdays."

"Usually," said Chucky. "Why?"

"I have a problem that needs attention. I'll do it on Saturday. See you then." Baca stood and flipped them a little half salute. He started turning for the door but stopped and turned back. "Oh, I almost forgot. I paid your little friend, Arn a visit at the jail. He seems to be perfectly happy. He found a chess partner when they are allowed out of their cells. He is working on trying to visualize the chessboard well enough so they can play without a board. Then they can shout their moves back and forth. He's also become addicted to sudoku when in his cell. Looks as if he's getting what he wanted." Baca didn't wait for any comment but spun on his heel and headed for the door.

When the deputy was out the door, Merv looked at Chucky with a perplexed expression. "What was that all about?"

Chucky shrugged. "He may figure the warrants for our arrest will be ready by then. It would be easier to nab us here when we are away from any heavy artillery."

"Yeah, they could just back up a truck with a power tailgate to make everything easier."

The next morning, Chucky was already in his place when Merv arrived. As Merv was dealing with his bench, Chucky was yawning broadly.

"You look like hell," said Merv.

"I never did get to bed last night."

"Why?"

"I finally started putting some things together. A friend e-mailed me some political stuff. There was one comment that got me thinking. This was the one that caught my eye." Chucky slid a sheet of paper down the table. Merv picked up the sheet of paper and read the printed matter aloud. "There are a lot of bad guys walking around because it's against the law to kill them."

"I started thinking," said Chucky. "What would happen to all those bad guys if breaking the law didn't matter to a group who wasn't afraid of the consequences."

"Tell me more. What kind of group are you thinking about?"

"I'm patterning my thoughts around our roundtable guys who already have no future. I'm calling this category of guys, 'A Few Good Old Men'. Actually, what I am visualizing is not an organized group, but a collection of individuals who espouse a common cause but remain unknown to each other. That way they won't face the association problems that we have."

"How would these 'Few Good Old Men' come up with their Final Mission?"

"Oh, there are plenty of injustices that should be addressed. Each man should choose his own mission. That way, part of the choosing process is to find one in a geographic area and situation where it is doable."

"I know a lot of people, who would do nothing without a strong support system."

"Then maybe those types shouldn't consider this path."

"Yeah, the ones I'm thinking about probably couldn't pull it off anyway. I take it that you're going to try exporting this concept."

"Anything to save the country," said Chucky with a rueful laugh.

"How do you plan to spread the word?"

"I'll start out by creating a blog and advise all those guys who have been e-mailing the post looking for information to go to the 'afewgoodoldmen.com' website. I'll ask them to pass it along.

"There are lots of dueling political blogs out there. Many have comment boxes. We can make some remarks along with the blog address."

"You're including me with this?" said Merv.

"Only if you want a piece."

"Yes, I want in, but I don't want to worm my way into what may be visualized as a sole proprietorship."

"There will be more than enough work to go around. Of course, there could be a lot of grief if something goes wrong."

"I already know about grief."

When Saturday arrived, Chucky and Merv would rather have stayed on their computers spreading the word about www.afew-goodoldman.com, but Baca had in essence made an appointment with them. They had been apprehensive as to Baca's reason for coming. The suggestion about arresting them had been made half humorously. Now they weren't sure after pondering the situation.

Chucky and Merv were in their usual spots when Lieutenant Baca appeared at the door with an elderly man in tow, who was leaning heavily on a stout cane.

"Do you suppose that's Baca's father?" said Merv.

"I think you're right. I can see several physical similarities."

Baca pulled out a chair for the old man. "Hi, guys. I'd like you to meet my father, Ladoslav Baca...Lad for short. Dad, that's Chucky and this is Merv," said the deputy, who bobbed his head in the appropriate directions.

"I'll be back in about half an hour. Dad wants to ask some questions and I probably don't want to hear the answers." Baca departed.

Lad Baca shook his head. "It's weird living around a cop. He has a very narrow view of the world. I spent most of my life in the Army and I have a difficult time following his restricted view of a broader scene." Baca senior waived for a beer.

"He seems to be a very astute police officer," said Chucky.

"Oh, he's a good cop, but his instincts don't go far beyond the county lines. His various capacities, or lack of them, are not why I'm here. I want to inquire into your views on revenge."

"Revenge?" said Merv. "

"Revenge, retribution...whatever you want to call it...and how you tied it into the 'endgame'."

"I'm not sure what you're driving at," said Chucky.

"My son, Bo, has pretty much figured out what's been going on around here. And there has been a lot of interesting chatter speculating on the various activities that all seen to emanate from this table."

Baca paused for a moment, but when no one commented, he continued. "From what I think I know, short-timers are exacting some sort of revenge on wrongdoers. The current theory is that you guys exchanged revenge accounts and once the pound of flesh has been extracted, the avenger chose to go down in flames...if you'll pardon my clichés.

"There are still some blank spots in my theory. I would guess that you guys worked on the barter system. But, why concern yourselves with old grievances? If they were that important, why weren't they taken care of years ago? And what inducement is offered to get an old man to travel to faraway places on a one-way ticket? Another question is how does one find actions to avenge?" Again Lad Baca stopped to see if he could get some feedback.

Chucky adjusted his position in his chair. "It seems to me that you have been doing too much thinking, making everything complicated. You don't have to find someone with a grievance. Wrongdoers are all around. Today there is little or no accountability. I was brought up to believe that bad actions should not go unanswered. The world is full of bad actors since so many have gotten away with bad actions.

"There are illegal activities all over the place. Look at the fraud in all levels of government...how about Medicare/Medicaid. Every industry, business, union has its own breed of corruption. Take your pick. There is no shortage of wrongdoers."

Labs Baca stared through Chucky for some time. "I had not projected this up to those scales. I was stuck on a mutual back scratching theme."

"That plan might have hazards," said Chucky.

"Yeah, that could cause someone grief," allowed Lad. "I still don't see any incentive to take on what could be a terminal project. I know I'd rather watch TV than tramp around the country trying to right an old grievance."

"When I was in Bosnia-Herzegovina, we were working in some rugged mountains. There was a good sniper dogging our trail picking off our guys with too great a frequency. We never were able to track him down or get air support at the right time and place. One day he knocked Grady down. It was a bad hit. It was foul weather and the chances of getting Grady out in time were bad to nil. We were hunkered down in brush and rocks. Grady had us watch the next ridge and he dragged himself up and took another shot so we could spot the sniper."

Lad Baca said nothing for a bit. Quietly he asked, "Grady?"

Chucky shook his head.

"The sniper?"

Chucky's head nodded.

After Chucky regained control of his voice, he said, "Sometimes, when there is no tomorrow or tomorrow is not worth living, a worthy Final Mission might have all sorts of appeal. A Few Good Old Men might make a world of difference in this crazy country."

Merv injected a comment, "Be sure to take note of the word 'worthy'." One wouldn't want to disgrace himself and any he leaves behind. Your son just pulled into the parking lot."

"Already? Don't tell him what we talked about. He wasn't too keen on bringing me here. I think he's afraid I might get involved with your conspiracy and if there is any legal action taken, I'd get scooped up too. That wouldn't help his career."

"You ready to go, Dad?"

"Sit down and have a beer. I hardly had time to finish one and I'd like another. We were swapping war stories. Chucky was in Bosnia-Herzegovina. You used to fly in and out of there, didn't you?"

The deputy reluctantly sat down and ordered a round from Babs. To carry on the conversation, he acknowledged that he occasionally flew from Rammstein Hospital to ferry in medical supplies and equipment. Everyone contributed a story before Baca could get his father to move.

After the pair left, Merv said, "What do you suppose that was all about?"

"I think it's a battle of wills. Both are alpha males...hard heads. The old man was always top dog. Now his son has taken over. But, pop still wants to show he has relevance and he can still impose his will."

CHAPTER 19

Chucky and Merv activated their website, "AFewGoodOld-Men." It was quickly becoming much more work than either had envisioned. Besides having to write and post meaty material, editing comments sufficiently to keep themselves out of trouble with their Internet host and the law was a constant drain on their time. New emails kept the computer dinging until Chucky had to turn off the sound.

There was no way they could respond to the e-mail traffic, so they addressed similar questions in little sidebars.

They still took their daily beer-break at the legion where they were fielding even more questions since the website went up. However, they now had a well considered message.

During a lull in the inquisition stream, Merv made an observation. "We're passing the word to a lot of people but as far as I can tell we have no disciples."

"This is the type of thing that doesn't lend itself to easy decisions. Probably a lot of anguished thinking is going on. Also, many will be struggling with moral dilemmas. There may be events out there that don't appear to link up."

Ex-military guys," said Merv, "should have a running start. Many have already waged those battles. I know I have. You?"

"Yeah. I was lucky when I hit the battlefield. I didn't have to fight a religious war too."

Babs, who always kept an eye on everyone approaching, sounded off. "Hey Chucky. Here comes your deputy."

Baca dropped into a chair across from the two legionnaires. Without preamble, he said, "What did you do to my old man?"

"What are you talking about?" said Chucky as he glanced quizzically at Merv, who shrugged. "We didn't do anything to him."

"Something's going on. He moped around for five days. He didn't get into any arguments or even complain about his arthritis. The next thing I know he's all smiles and good natured. He spent hours in his garage, which has a small workshop. He wouldn't tell me what's going on or what he's working on in the shop."

"Go out to the garage and take a look," suggested Merv.

"Oh, he doesn't live with me. He has a small, older condo that was built back when they had garages. I'd have to break in and that could cause all sorts of problems. Did you give him something to avenge?"

"Not us," said Chucky." We have enough problems. We don't need any one trying to link us up with any revenge cases."

"You've read our blog, so you know what we are advocating," said Merv.

"Something's going on," growled Baca. "If you've enticed him into your little vendetta club, I'll have your heads impaled on lances in front of this place."

Chucky laughed. "And where would you come up with lances?"

"There are two, crossed cavalry lances on my dad's living room wall. Since he retired, he's been working in metals. He fashions all sorts of old weapons and gadgets. You know what I mean?"

After Baca left, Merv said, "What do you suppose the old man's doing?"

"I'd say he's found a mission that pleases him and his son

knows it. The deputy is covering his ass by announcing to everyone he doesn't know what's happened to his father."

"You suppose?"

"Yeah. I think the deputy hates to see his father decline into old age and pain and he's secretly looking at a Final Mission proposition as a worthy endeavor."

"If the old man has found a mission, how do you think the deputy will take it?"

"I'd say that all depends on the mission and how it comes out. Baca will take care of himself."

****

Because of the broken levels at Merv's apartment, Chucky had problems negotiating the terrain. So they ended up adding a desk and one of Merv's reinforced chairs to Chucky's small apartment. That way they were able to talk things out before establishing their stand on a new subject. They were still trying to stay rather vague because there were a lot of social and legal ramifications to the proposition that they were fostering.

They dropped into the routine of having private time until they met at the legion for their beer. The commander put a box beside the bar into which he tossed the e-mails that still came to the post. The guys picked those up and fielded questions from visitors.

The character of the e-mails both to the post and to the blog started to change. More and more links to articles and attachments of scanned newspaper clippings of both lethal and nonlethal events scattered around the country. Chucky put out a subtle plea over the blog not to send such information to them as it might bring unwanted attention to such events.

After the stand at the legion, they had lunch...sometimes together, sometimes solo. After lunch, work began at Chucky's. As time passed, they were finding they were working deeper and deeper into the dark time of the days.

One day, deputy Baca was waiting for them at the legion. "Dad disappeared. You have any idea where he went? He left sometime yesterday or the night before. His car is gone. He left a note telling me to mind my own business and he'd pop up when he was ready."

"No," said Chucky. "The only connection we had with him is when you brought him here. I doubt if he had any program in mind when he talked with us."

"With your resources, you could probably find him," said Merv. "Are you?"

"No," said Baca quickly enough and with sufficient force to make it apparent that he already rejected that approach. "I just have to hope for the best." Baca departed with no further comment.

"What do you suppose the 'best' will be?" said Merv.

Chucky didn't speak for a moment. "I have a distinct impression that Deputy Baca feels the 'best' would be that his father succeeds in a Final Mission."

Deputy Baca didn't have long to wait.

CHAPTER 20

The time spent with Chucky and Merv had not answered any questions for Ladoslav Baca. It had provided him with some guidelines. He needed a cause that the majority of people would find worthy. If the endeavor was found to be wanting, then so would he. The taint of an injustice could spread to Bolek, which was unacceptable.

Ladoslav Baca was very much a political entity. While in the military, he had curbed his propensity to express his feelings. After retirement, he wore his conservative philosophy on his car bumper and his back in the form of outlandish T-shirts.

Then one night, he had what he considered a brilliant idea while watching TV news. The rest of the night was spent in creating a plan of action. Also he had to design the necessary equipment.

After his decisions had been made and a course of action plotted, he was pleasantly surprised that getting out of bed in the morning wasn't nearly as arduous a task as normal. The whole world looked a little brighter. He couldn't even find a reason to bark at Bo.

His days were busy with shop work and his nights were filled with research. Finally, everything was in order. The next day his mission would begin.

Ladoslav had to take into consideration his reduced physical capabilities when he calculated his driving time to Atlanta. In his younger days, he'd have driven straight through. However, he couldn't sit very long anymore. But, time was not really that much of a problem. The Senate cloture vote on the immigration bill wasn't until next week.

Before selecting a motel, Baca found the offices of Liston and Liston, Attorneys at Law. It turned out to be an elegant old house that had been converted into lavish law offices.

His reconnaissance had paid off. The area was too highbrow for his old car. In addition, the parking was located behind the building and that was too long a walk for his old legs. He'd take a taxi, which could drop him off under the portico, at the front door.

After circling the block a couple times, Ladoslav found a parking spot across the street from which he could watch the office activities. Nothing of importance attracted his attention. As the afternoon lengthened, finding a motel became the next item on the list. He wanted to make an appointment with Larami Liston, the son of Senator Lansing Liston.

The Liston phone was answered by a real person, who courteously transferred his call to the appropriate person to make an appointment.

"General Baca, Mr. Liston has a very tight schedule. I could have one of his associates see you sooner."

"No, Senator Liston, told me some time back, that when I was ready, I should bring the document to either him or his son. They could handle it. I only need to see him for a few moments to get a receipt. I see from the TV that the Senator is tied up in Washington so the son will have to do."

"If you can be here just before 10:00 in the morning, I can get him to see you just before his 10 o'clock."

"Thank you, Miss."

Baca paid the taxi to wait across the street from the law offices until eight minutes before the hour. It took him three minutes to painfully make his way to the receptionist. The long drive had caused a real flare up. He leaned heavily on a stout ornate cane as he made his way to the receptionist desk.

The young lady said, "General?"

Baca bobbed his head.

"Yesterday we failed to get your full name. We couldn't pull your file."

"It's Ladoslav Baca." Baca spelled it in the Slavic manner.

"Oh, we probably had the spelling wrong."

"There shouldn't be any file on this. It was strictly a private conversation."

"Please wait for a moment and I'll see if Mr. Liston is available." She bounced up and entered the first door on the left off the central corridor.

Baca leaned on his cane, while the receptionist was in Liston's office.

The front door was opened by a business type in a suit, stretched over a pot belly.

The girl reappeared. On seeing the recent arrival, she smiled brightly and said, "Good Morning, Mr. Swenson. Have a seat. Mr. Liston will be available shortly." Turning to Baca, she said, "Go on in. Mr. Liston will see you."

Baca picked up the ordinary folders that normally contained a legal pad that he had placed on the desk and hobbled to the door. When Baca made it through the door, Liston stood and walked around the desk to greet his client.

"General. How nice to meet you." The two men barely touched hands. The attorney was respectful of the gnarled arthritic hand that was presented.

"Please, have a seat and tell me how I can help you."

Baca settled into a comfortable chair in front of the desk. Leaning his cane against his leg, he fingered the folder in his lap.

"I have a document concerning a cache of German gold." Baca started to stand as he held the folder forward. The cane slipped along his leg before falling to the floor. Baca leaned forward to retrieve the walking stick. While hidden under the desk he held a little button down and twisted the ornate handle.

When he stood to offer the folder to the attorney with his left hand, it slipped out of Baca's old deformed hand. He lurched forward to catch it as did the attorney. Simultaneously, the old man jabbed the cane across the desk into the chest above the heart. There was a loud explosion. Liston recoiled backward into the wall below the bay window. His chair tipped over, dumping the dead body onto the floor.

Baca immediately opened the folder and laid out computer-generated show card on top of the desk.

A scream and the sounds of movement came from the outer office.

Baca removed the cap on the cane handle. As the office door was flung open, Baca placed the tip of the cane on the carpet and threw the weight of his upper torso against the open handle. Another sharp report sounded and Ladoslav Baca's lifeless body slumped to the floor.

CHAPTER 21

The word spread quickly. The bold murder of a prominent citizen and the subsequent suicide by the murderer is big news. All the news agencies were on it instantly. There would be even greater interest because the victim was the son of the senior Democratic senator from Georgia.

Atlantic City police sealed off the scene and they were very stingy with information. Details would remain sketchy until the Senator would permit their release. Reporters were really going to have to dig. Politicos seldom were forthcoming with this type of material unless there was a distinct benefit to be gained.

At first, the identity of the assailant was not revealed. Apparently, the police matched him by the ID he carried, but that information was not forthcoming.

However, as soon as the morning mail delivery hit the various newsrooms at TV and radio stations and area newspapers, the word was out. They all received copies of the material that Baca had left in Liston's office.

The mailing said, "By now, everyone should know the name of Ladoslav Baca. I am one of the "Few Good Old Men," who has just completed my Final Mission. The Dems are buying their cloture vote on the Senate immigration bill. However, how many Dems are now willing to vote this monstrosity into law? The Senators who vote for this bill have more family members and friends than can be protected. I'm sure many Good Old Men are looking for a worthy Final Mission."

Deputy Bo Baca received a call from a friend who had heard the TV news alert. When Baca received the call, Chucky's American Legion Post was the closest known TV set. He broke a few laws getting there. No one answered his first ring. He impatiently jabbed the bill for two prolonged rings.

Begrudgingly, a patron whom Baca didn't recognize answered the door, but then blocked the way saying, "What cha' want?"

"I'm a deputy sheriff here to see Chucky." Baca didn't want to give his name at that moment.

When the two men passed through the inner door, Baca noticed that all TV screens were tuned onto news stations. It had been Babs who made the name connection and alerted everyone to the unfolding event. She was also in charges of the remotes, so she could bring up the volume when the Baca story was on and mute for the commercials.

All eyes were on the screens. Babs recognized Bo but honored his headshake. Chucky and Merv were watching the TV over the bingo board, so they were looking the other way when Bo came in. They became aware of him when he dropped into a chair across the table. He waved off all comments until the station went to commercials.

"What have you heard? A friend called as soon as he heard the name, but he didn't know what had happened."

"There aren't many details yet," said Chucky. Earlier there was an announcement that Larami Liston, son of Senator Lancing Liston had been murdered and then the murderer had committed suicide. No information concerning the event was being released. They knew the victim was Liston because his office called 911 and reported he was down. The EM guys returned empty-handed and homicide arrived. The office staff said Liston had been in his office with an old man, who was not a regular client.

"From that point on, there was a tight clamp put on all information."

"It would appear," said Merv, "that the police are trying to sit on everything, probably on the Senator's demand. But your dad foiled part of that by sending all the media a statement identifying himself and saying that this was his Final Mission."

"How'd he do it?" said the grim-faced deputy.

"That info hasn't been released."

"Did you guys know anything about this?"

"No," said Chucky. "We haven't had any contact with him since you brought him here. We're sorry for your loss."

Bo Baca shifted his gaze back and forth between the two men. "I don't know if I should rejoice or cry. I'll probably cry for loss of my father, but at the same time I'll rejoice for his freedom from that old worn out body and a future of perpetual pain and a disposition that goes with it."

Another update came across the cable channel. Everyone stopped to listen. There was little additional information concerning the event. The big news was that Senator Liston was already in route aboard a Navy jet. He issued a statement condemning right wing radicals who now had demonstrated that they constituted a clear and present danger to the US government. They are now attacking the United States Senate.

"Bah," said Baca, "dad was apolitical. He was for whoever approved the military budget. One thing that really wound his clock was illegal immigration. Our grandparents did it the right way. That Senate bill, which is designated to give amnesty to all those wetbacks, really frosted him. He was unhappy with both parties for their failure to take care of the problem."

Chucky watch Baca closely as he said, "You know the authorities will turn up here shortly."

After Baca departed, Merv said, "Do you have anything to clean up?"

"No. I have a substantially electronic trail and if I reformat my hard drive they'll accuse me of tampering with evidence. The only thing I want is to let Arn know what's going on and to check 'The 'Trucker's' status."Tomorrow's visiting day. I think I'll go see Arn while I have a chance. Want to come along?

"I'd better pass. Those jails aren't set up for the likes of me. I can't stand that long and there's probably no place that I can sit. Give him my regards."

Chucky had to go through the usual rigmarole before he could take his chair into the visiting room. They patted him down and minutely searched his wheelchair, even though there would be a glass panel between him and the prisoner.

Arn came hobbling in using a walker. "I'm trying to trade this steed for one like yours," said Arn with a bright smile.

"How are things going?" said Chucky.

"This is better than the old folks home where I used to visit my mom before she died. This place is a lot cleaner. The furniture is newer. While the food isn't great, it's better than mom used to get and it's all free."

"I wouldn't call it free, but I guess that's a matter of opinion."

"Say, my landlord came by and I told him I'd have you send him a check for a month's rent. Will you call Chase Bank and have them send me a signature card on my checking account? I'll have you put on my account. I have an account here in jail. Send $100 a month to me and donate the rest to the legion. Send some of the young guys to clean up my apartment. There isn't much. Guys can take whatever they want and throw the rest into the dumpster."

"We can do that."

"How's Merv?"

"I think he hurts all the time but he's able to cover it for the little bit that he is at the legion. Since we started the' afewgood-oldmen blog, we're together much more and some of the pain shows through. He would have come down with me but he was afraid there would have been nowhere to sit.

Chucky proceeded to tell Arn of Merv's experience with the police in Venice. Then he switched back to serious matters. "There will probably be some federal types lurking about shortly." Chucky gave a very abbreviated account of Ladoslav Baca's activities. "Merv and I will be linked through our website and the roundtable. You don't have anything to worry about here. You are out of circulation before any of this came down. But because of Senator Lipton, all the trees will be shaken."

"It'll just be an interesting interlude as long as it doesn't interfere with my chess."

On the third day after the shooting, two FBI agents showed up at the legion. They were waiting at the round table when Chucky arrived. Chairs had been removed from the table to form a conversation circle in the corner as far away from the bar as possible. Merv's bench was there along with a spot for Chucky's wheelchair.

Babs gave Chucky a head sign as he came in the door. He wheeled himself to the service gate at the bar to converse with Babs. One of the agents bounded up to intervene. "I'm Agent Baker from the FBI." He flashed a badge momentarily.

"Oops, not that fast," said Chucky and held out his hand for the badge holder. The agent wouldn't surrender it, but did hold it out for close scrutiny.

Satisfied, Chucky said, "What can I do for you, Agent Baker?"

My partner and I would like a word with you and Mr. Mervyn Dawson. We set up a little area in the corner for as much privacy as we can get."

Merv's car pulled in the parking lot. The other agent positioned himself near the door to intercept the new arrival.

The other agent identified himself as Agent Adam, which brought smiles to the faces of the pair of legionnaires.

Baker started off the session. "We're here just to have a conversation. You undoubtedly know the reason for the visit. The murder of a powerful Senator's son is bound to bring great tension into the lives of anyone remotely connected to the event.

This includes you two because the Senator is convinced that you are the proximate cause of Larami's death. Your hateful, radical, right wing blog influenced a poor, addled elder to commit a heinous act against an innocent unsuspecting citizen.

"We are not even supposed to interview you since all of the evidence the Senator feels he needs is already part of the public record. You'll just have to sit by and see how this plays out. Don't leave the area."

Adam added his thoughts. "It might be helpful if you voluntarily shut down your blog. If it continues, the Senator will probably bring pressure to bear to have it removed."

"We won't be at a loss for something to think about this afternoon," said Chucky rather caustically. "How do they say it? We'll have to take your warning under advisement."

The agents were not going to argue. They had carried the message as they had been instructed to do. The assignment went contrary to their normal instincts. They didn't like being the bearer of threats from an old blowhard politician, who had become an expert in throwing his considerable weight around.

The agents didn't make any pretense of rearranging the furniture. They just marched out the door without uttering another word.

Chucky and Merv had to deal with Babs and the commander before they could settle down to discuss the latest developments.

"You think the Senator will go after conspiracy charges?" said Merv.

"Probably. He can't do anything to old man Baca. I imagine his son was smart enough to remove himself for consideration, so he has us shouting 'FIRE' in a crowded theater. It will be a while before they can build a case. We have to formulate a plan so that the movement can survive. I've come to believe that it's a worthy cause."

"I'll agree with you there. You have any ideas?"

Chucky didn't answer right away. "We have to become a hydra."

"A hydra?"

"On a regular snake, if its head is cut off, the creature dies. With a hydra, there has to be a lot of chopping before the body quits wiggling. We should develop a bunch of subsidiary sites that can take over. Give this some thought."

CHAPTER 22

With a heightened sense of urgency, Chucky and Merv started building a diverse network on the Internet. They didn't abandon their principal blog as the FBI agents had suggested, but neither did they get too provocative.

Sorting through back emails, persistent followers were separated into a new file. Bloggers of a similar point of view were found and encouraging comments were sent.

The long hours were taking a toll on Merv. He was working with total conviction. Chucky speculated that Merv was in persistent pain and he was working with such dedication because it diverted his attention from his physical problems, but he was wearing down.

Chucky started closing up shop at 10pm ostensibly so that he could go to bed. Actually, he was sending Merv home to get some rest. Also, he had another solo construction project in his little workshop that was calling for attention.

As the days passed, Chucky and Merv and all the legionnaires paid attention to the news coming out of Atlanta. They charted the activities of the Senator and tried to analyze his periodic rantings.

This all became background noise with the news about another supporter of the immigration bill had lost his wife in a savage automobile accident by another self proclaimed "Good Old Man."

Three days later, another senator lost his big, fancy house to a fiery crash. An old man had driven a car with a propane tank strapped to the front bumper, into the side of the house. They hadn't figured out what type of igniter had been used.

One morning Chucky was late getting to the legion. As an explanation, he said, "Sorry, I got involved in a discussion of the Immigration Bill. There has been a projected date to have both the House and the Senate bills passed, but there is so much infighting that the majority can't be certain of enough votes.

"At one time they thought it would sail through without a major challenge. The Republicans can't stop it. The Dems are getting a little hesitant because the new polls show overwhelming public disapproval, which until now, they have ignored.

"From what I've heard," said Merv, "is that it will eventually pass because there are enough votes that can be bought. It's just taking a little time to establish the price."

"I think there is another factor in play."

"What?"

"Not what, but who."

"Okay, who?"

"Ladoslav Baca."

"Baca?"

"Yes. His Final Mission was a protest over the Immigration Bill. Since he picked off Liston with his bang stick, we've had a congressman lose a wife, and a senator had his house destroyed because of their political stance. I think Baca has given the great, unwashed masses a weapon to express themselves.

"As it stands now, their only real protest is a 'No' vote in some distant election...after the damage is done."

Chucky waived for another beer. "Now their only voice is to

show their objection in the polls, write letters to the editor, or their congressmen and yell at town hall meetings. Lip service is paid to listening to this feedback, but I've never noticed much change of thought based on all of that expended energy."

"So far," said Merv, "there hasn't been any concrete evidence that any political minds have been changed. I'd say that Senators are starting to watch in the rearview mirrors, but they'll continue to operate in their own self-interest."

"I think you're right. But this is a new phenomenon on the scene and unless you're an internet junkie, you probably wouldn't have connected any attacks. The media is ignoring it. So I came up with an idea to enhance the public's voice in running the country."

"Oh, tell me more.

"I think it would be interesting to start a 'Hit List' for all those who vote for the Immigration Bill."

"A 'Hit List'. Isn't that getting a little extreme?"

"This list isn't necessarily a lethal list. Hits can come in a thousand different forms. But first, let me describe the list.

"The basic list is the name of every Senator or Representative who takes a position to advance the bill, which is obviously contrary to the wishes of the people. Under the name of each are vital statistics, such as addresses, family members and their addresses, schools, places of employment, churches, businesses, favorite restaurants. To obtain this data we should use Wikipedia's formula of requesting the public to supply information. It would be up to us to make the entries into the blog."

"What is the public to do with all this information?"

"If the bill moves through its various stages, it would be the public's job to make life miserable for all those who voted in the affirmative. This will be amplified if those who have any connection with the target are also subject to the wrath of their constituents."

"All those other people probably had nothing to do with that politician's vote," said Merv. "You'd bring the wrath of the world down on their heads too?"

"Yep. Most of those political types show little or no concern for those they represent when they vote contrary to the stated will of the majority. They seem more concerned with their own house-keeping. It takes a bundle to keep that bunch happy. Besides the pay, look at the more than generous health care, retirement, travel, office expenses and so forth. Then they sell their souls to special interest groups and top it off by selling their vote on major bills.

"Take this immigration bill for example. In its current form, it's going to heap additional expenses on businesses, hardships on hosts of people, taxes on our social services' budgets and will expand bureaucracies. And the beat goes on. The people affected by a 'yes' vote didn't do anything to warrant it. And as it stands, they are not going to have even a voice in the formation of the bill.

"As soon as the immigration bill is out of the way, the energy bill will be the next move toward socialism."

"Yeh, something has to be done," conceded Merv. "That socialism creep is moving too fast. By the next general elections, our freedoms could be history. We have to be careful not to promote an illegal act.

"The term, 'Hit List' could easily be misconstrued as a call to kill all of those folks, which could be serious enough for them to come after us."

Chucky gazed off into the distance for a bit. "You're probably right. We shouldn't give them an excuse to come after us."

Chucky started to laugh.

"What's tickling you?"

"How about calling it the 'Bird List'?"

"Bird?"

"Bird, like in 'finger'."

"Oh," said Merv with a chuckle.

"In the blog, expand on 'finger' to mean any form of expression appropriate to convey one's feeling concerning an affirmative vote."

"Yeh. That should be broad enough and it doesn't call for any physical harm."

"As we go along," said Chucky, "we can report on any more dramatic expressions. I imagine it won't be long until more creative approaches appear.

"I'll set up a new blog this evening. We can download all the names of the Senate and the House from the government website. There is a lot of other information available. That will give us a good start.

CHAPTER 23

It took a couple of days to get the blog up and running and pass the word around thru afewgoodoldmen. It seemed to have hit a note of accord with a whole different group.

When Merv arrived at Chucky's in the afternoon, he was told to take a look at the email for the Bird Blog.

"Wow, three hundred and fifteen."

"The majority contain information concerning family, residences and business addresses. There is a lot of duplication. It's going to take a whale of a lot of work to sort through this."

"We already have as much work as we can or want to handle," objected Merv.

"It would appear we have a valid idea. Now we need to outsource this work."

Merv shook his head. "That might not be easy to do. Our Bird List has both Democrats and Republicans on it, so anyone with a strong party affiliation wouldn't be inclined to further this enterprise. Libertarians and general run of independents aren't organized enough to handle the volume of work that this might entail and there is a pretty high degree of danger involved."

Chucky didn't say anything for a bit. "What do you know about the 'Mad Hatters'?"

"Not too much. They're ignored by the mainstream media, which passes them off as ignorant trailer trash. There's more to them than that. They are beating the drum for eliminating corruption and stopping waste. They fear for the economic future of the country. They are deficit watchers."

"From what I can tell," said Chucky, "they are getting better organized than most of the other splinter groups."

There are a lot of business and pro-types in the group. They wear those funky top hats so they can carry a front sign and a back sign stuck in the band around the crown. Those people are not much for sign waiving."

In two more days the email count had expanded to 800 plus. The next day it was zero.

"Those bastards pulled the plug," cried Merv, when he tried to call up "the Bird List."

"We knew they would shut it down sooner or later. Although, I didn't think it would be this soon. Our meeting with the Mad Hatters isn't until tomorrow. Tonight let's download all the bird list sites and burn them on DVDs. Tomorrow night, the Mad Hatters should be able to get bird list #2 up and running. Then everything will be up to them and we can get back to our own work."

"That's all right," said Merv. I'm ready for a change."

"I think things are already changing," said Chucky. "I have a very distinct feeling that somebody has been rummaging around in my place. I set out a few trips to let me know if I have uninvited guests. They would have to visit while we're at the legion.

"Your place would also be available for illegal entry while you're over at my place in the evenings.

"Come to think about it, somebody may have already visited

my place. I habitually unplug my computer when I leave because of all the lightning we have. A couple of days ago it was plugged in when I checked my mail in the morning. I dismissed it as forgetfulness. Maybe it wasn't."

"If somebody has been messing around," said Chucky, "they were probably copying our address books and emails. I don't recall any traffic between us and either of the new attackers, who have been reported on the internet."

"I doubt if I would remember just email contacts." Merv shifted to ease some discomfiture. "Who do you think may be poking around in our stuff?"

"It would have to be federal. I don't think the locals would dare. State has no interest, but Lipton could exert enough pressure to get somebody to break the law."

"According to the TV this morning, the House and Senate claim they're close to a compromise bill on immigration. Lipton will have to get back to DC pretty soon. The progressives are going to have to buy some votes or they'll never get past a filibuster."

CHAPTER 24

It had been a busy evening. Chucky and Merv had a beer and watched the late TV news before Merv started for home. Merv levered his bulk onto his feet with a grimace. He made his way to the door. "Goodnight. See you tomorrow."

Just as Merv turned the knob and started to open the front door, a large body slammed into it from the outside, knocking the knob out of Merv's hand. The door was pinned against the wall as a black-clad figure ricocheted off it into the room, brushing Merv, who lost his balance and went down in a heap.

A second figure charged into the room pointing a semi-automatic with an ugly suppressor at Chucky's chest. Four more men filed into the room.

The final one was obviously in charge. He began barking instructions. To Merv he said, "Stay put." To Chucky, it was, "Don't try to be a hero." To the two with drawn guns, he nodded toward Merv and Chucky. They took up a position behind each of the legionnaires. Two other invaders each took a computer and began hooking devices up to the USB ports.

Chucky's guard conducted a quick, but thorough pat down and looked into the various pouches of the wheelchair.

With well displayed disgust, Merv's guard patted down those areas he could get to. Merv groaned.

The leader sent the sixth man into the kitchen.

Chucky hadn't said a word other than an expletive on the initial break-in. Now he turned his chair slightly so he wouldn't have to get a kink in his neck to face the leader. The action brought a stern warning from his personal guard, which Chucky ignored.

Merv was trying to readjust himself but his guard wouldn't let him.

Chucky addressed the boss. "Let my friend adjust his weight. You don't want an autopsy to find out he was dead before the fire."

"What are you talking about?" snapped the leader.

"That weight has to be distributed right or he can suffocate or cut off the blood supply to a vital organ. It's not as if he's going to jump up and sprint out the door before you can bat an eye."

Merv's guard was given a nod. While Merv struggled to lean against the wall, Chucky took the opportunity to inspect the invaders. They were identically dressed all in black. There were no insignias. All the men were of a basic type...large, well muscled and obviously fit. No pretense was being made to obscure their faces.

Chucky could hear the man in the kitchen doing something to the gas range.

The man copying Chucky's hard drive and external drive said, "This will take a while. There's a lot here."

"We're not on a schedule."

Merv hauled himself into a sitting position so that he could lean against the wall. He was breathing hard from the exertion but he still gasped, "Thanks" to Chucky.

While regaining his breath, Merv was taking in the various scenarios that were unfolding around the room.

"Why don't they just take the hard drives instead of spending all his time copying?"

"In case the computers are not destroyed in the fire, it would be suspicious to either have no hard drives or brand new blank ones. It would no longer look like a regrettable accident. That's why our buddies here won't use those guns that they are pointing at us. Bullet holes would be hard to explain."

"What made you so smart?" said the leader in an ice cold voice.

"Bosnia-Herzegovina. Both sides did a lot of booby-trapping. You learn the ropes or you won't last very long."

Bosnia?"

"Yep. Who sent you?"

"Hah."

"I guess that's all right. If you gave me your acronym I probably wouldn't recognize it because it's buried so deep, only a few people even know it exists."

"You are a smartass, aren't you? said the leader. Why don't you tell me what we're going to do."

"You just pulled off an armed house invasion but you're not concerned that we are getting good looks at you, which means we're not going to be around to identify you. All of you are leaving fingerprints and DNA all over the place so there must be a fire. I'm just wondering how you're going to trigger the big propane tank right beside the apartment without leaving any evidence."

"I'm leaving that to Ernie. He's the expert."

Chucky's gaze swung back to Merv, who was breathing better. "How you doing buddy? Do you suppose this qualifies as a Final Mission?"

"I couldn't have picked a better one."

The leader became alert. "What's going on here?"

Chucky swung around to face the leader. "You've already lost."

"Bullshit."

"See the camcorder up on the book shelf?"

Concern showed on the invaders faces. As they all scanned the shelves for the camera, Chucky jerked the left armrest of his wheelchair back and rotated it sharply to the left.

A powerful explosion ripped the interior of Chucky's tiny apartment to pieces instantly killing all those inside. The only survivor was the tech who had gone outside to work on the gas tank and he was badly injured when a block wall smashed into him.

When the first 911 call came in, a sheriff's office patrol car was in the near vicinity. Deputy Hal Martin was on the scene almost immediately.

One of the responding neighbors had found the injured tech. Martin called for an ambulance. A few small fires had been started but flammable items had been badly scattered so nothing became a problem.

The roof had collapsed on the remaining parts of the walls, making the apartment hazardous to enter. With his flashlight, he could see figures scattered about the room. When he spotted the mangled remains of a wheelchair and the mound of flesh that had been Merv, he recognized whose place it had been. Martin had been involved in some of the early inquiries at the legion. He had the dispatcher put in a call for a Bo Baca.

Lieutenant Baca arrived at the scene as a tow truck was stabilizing the roof so that it wouldn't cave in. A paramedic went in first and reported there were six bodies and there were numerous body parts from another victim or victims. There were no survivors.

Baca was a ranking officer on the scene for the moment. The

sheriff had been called and with as many victims, it was politic for him to show up. Baca called for the coroner and forensics. The remaining paramedics departed. He sent one of the deputies, who had just arrived, to the hospital to get all the information he could find on the injured man.

All sorts of alarms were going off in Baca's head. The original paramedics had cut a black, long-sleeved knit shirt off of the injured male and a black baseball cap was left behind. When he looked inside, he spotted what appeared to be a side arm with a silencer. Those were rare and unlawful. There were six too many people at Chucky's. The clothing and the gun suggested military involvement. If that was true, there was going to be hell to pay.

Normally, he would've stayed away from the crime scene until forensic finished but he collected his point-and-shoot Canon digital camera and made his way into the ruins. He had a while before anyone else would arrive. Martin was tied up in controlling an ever-enlarging crowd of spectators. His powerful flashlight illuminated the room sufficiently for the camera to focus. Baca took shots as fast as his Canon could recycle. He had a large chip and a freshly charged battery, of which he took maximum advantage. He covered the entire apartment from every angle available to him.

Without touching anything, he found guns, electronic equipment and a toolkit in the kitchen that didn't look as if it belonged there. The bedroom was a disaster, but didn't appear to related to the event. The only remarkable thing was the jungle gym that Chucky had constructed so he could move about without his wheelchair.

In a tiny back room there was a work area that appeared to be used for working fiberglass. In the distance he heard a siren. Baca took shots of the whole room from as many angles as he could before he went out through the back door, which had been blown off its hinges.

Before the sheriff came slithering to a stop, Baca was back in front with his camera in his pocket.

"What have we here, Baca?"

"From the looks of the van parked in front, the mangled wheel-chair and the prosthetics equipment inside, I'd say this apartment belonged to Chucky Bentley. I think that lopsided sedan behind the van correlates with the huge body inside. It's probably Mervin Dawson."

The sheriff looked puzzled.

"The probable founders of the Few Good Old Men movement."

"Oh, right."

"There are five other bodies plus an injured one that has been transported. They are all dressed in black. There is what looks like military weaponry including suppression equipment."

"Shit."

"If that's military, we have a real problem. I don't suppose those guys were there at this hour to give the reenlistment pitch."

The forensic team pulled up. The sheriff moved off to talk to them. More troops arrived. Baca had Martin relieved of crowd control. "Hal, wander around the neighborhood and find the vehicle or vehicles that brought those six men."

Baca slipped his camera into a pocket of his briefcase before assigning duties to the new arrivals. Then two TV trucks started to maneuver into position. The sheriff had moved into the backyard. Baca went to advise him of the media's arrival. The sheriff held an elective office, so Sheriff Benis cultivated all the goodwill he can muster.

"Sheriff, two TV trucks just pulled up. It'll take them a few minutes to set up."

"Good. Bud is inside getting some digital photographs so I can see what it's like without disturbing any evidence. After I have a look, I'll come out and give a preliminary statement."

As Baca was passing the word to the TV crews, Hal returned.

"I think I found them. There are two locked rental cars in the sports bar parking lot around the corner. They don't belong to any of the patrons. The office is trying to trace them."

"Good work. I'm going to put a guard on them until we find out."

Deputy Martin was watching the flashes going off inside the ruins. "Didn't you already take a bunch of photos?"

"Yes, but let them take the official ones. Don't mention mine. This case involves a Senator and maybe the military or federal agents. A few years back, a Senator's daughter was killed in an auto accident with her black boyfriend. You'd be amazed how soon motor vehicle records, police reports, and photos just disappeared. In case the official ones suddenly can't be found, I'll still have a set.

"I have a feeling this is going to become a sticky one. Keep records of everything you do...time and date. Protect yourself."

It was a long night. Forensic was working under a difficult and dangerous situation with lots of victims and materials to process.

When the crime scene crew finally finished, a contractor was called in to shore up the building. At dawn, the medical examiner started removing the bodies. The biggest problem was with Merv. They had to build a frame over him, tear down the wall and bring in a lift truck and a covered truck.

Baca stayed around to control the scene while wondering if a governmental agency would show up, but none came. The news from the scene was enough to raise open speculation of a possible federal involvement, yet no agency claimed its boys.

Once everyone left, except for the deputy preserving the scene, Baca went back in for his own minute examination. Everything of interest in the front of the house had been tagged and removed.

The workshop in the rear had been a question mark. Apparently, forensics had placed little or no value on the contents. It looked pretty much as when he had left it the first time. There was a shelf with four peculiar looking fiberglass mounds, about 5 inches high and 6 inches across the base. They were very smooth on the outside, but the hollowed Interior was rough and unfinished.

By poking around, Baca finally found all the equipment and materials needed to make a form, a negative plaster mold and fiberglass to make the positives. He also found a stack of white knit pouches to cover the fiberglass forms. There were draw-strings to tie them on the forms.

Suddenly he realized what he had. These were the stands that Chucky used to sit up straight. Without the hip and butt, Chucky would be sitting on his balls all the time and slumped over. Baca checked the old stands on the shelf. They were basically the same but small variations were apparent. When Chucky's hip hole changed, he made a new one that was more comfortable. They were just rigid shells that Chucky covered with a knit cas-ing for sanitation purposes.

From Chucky's house Baca went to where the two rental ve-hicles had been parked. Keys to them were found in the pockets of two of the black clad men. Fingerprints placed them in the vehicle. There was still some place where they had changed clothes. There was no indication of where that might be. Since they were not carrying any motel keys, the speculation was that it was not public accommodations.

Baca went home to get some sleep. When he went to the office that evening, he found no one still had shown any interest in the six black-clad men. The injured one was in no shape to talk. Fingerprints from all had been sent to the FBI. There was no record on anyone with those prints. The files had apparently been purged, but when? The car rental records produced noth-ing useful. The names led nowhere. Their credit card accounts were valid but led to a false front.

The investigation was getting stalled already. Baca had sus-pected that there would be problems when he saw those military types. After going through the collection of dead inquiries, he decided he would surreptitiously reproduce all of value for his own private file. There were some powerful forces at work.

Baca stayed late working on files and figuring directions of inquiries so that he was there when the call came in reporting that the injured home invader had mysteriously disappeared. The deputy who was guarding him was found asleep...apparently drugged.

No one ever found out how the guy was smuggled out. Baca had the feeling that it was an inside job but nothing was ever proven. With as much money as should be available, anything could be possible.

After catching up on his sleep, Baca invested considerable time in printing additional photographs taken at the scene. His photos were about the same as the official ones. One area of difference was his interest in the little back room. Baca had taken many more detailed shots. One item that caught his eye was what appeared to be a small flashlight with the lens off. It was tucked away on a low shelf. Chucky didn't have any high shelves.

There were enough items of interest to warrant another trip to the house. Although there had been no official determination as to the source of the blast, he had the distinct feeling that it was Chucky who detonated the charge His body had practically disintegrated. The damage radiated from where Chucky was probably sitting in front of his computer. If Chucky had put together a charge, it would probably have been in that storeroom.

If that was Chucky's work, where did he get the explosives? What kind were they? Why did he have a charge in his wheelchair? Did he know he was going to have unwanted company?

On the other hand, the charge might have been an offensive weapon. Chucky hadn't gone out on a Final Mission. Maybe he was getting ready to go.

Before going back to check out the little shop, Baca stopped by a friend's house. Dan was a dog handler for TSA. Juno was a golden retriever, trained to seek out explosives. Baca asked Dan if Juno could be around a blast scene without blowing out her nose. The trainer didn't think a short exposure would cause any problem. Besides, Dan wanted to get a closer look at a real blast site.

Baca led them around the other side of the structure so as to stay away from the blown out area. Juno didn't even get into the house before she hit on something in the lawn between the patio and the concrete walkway. Baca went into the shop for an old tire iron he'd seen in there. He pried up the corner of the sod and Juno became excited. Four c4 explosive wrappers were stuffed about 4 inches down. That much c4 would about fill one of those fiberglass stands Chucky used.

Another part of the puzzle was probably answered when the flashlight turned out to be a magneto light that built a charge by shaking a weight along a slide in the handle. The switch was also missing.

Baca had to report the c4 wrappers but he didn't report his speculation over the flashlight. This case was never going to be prosecuted. The trail would end with the five unidentified bodies and a missing person. The indications of high-powered involvement increased when the autopsy report came back. All five bodies were like peas out of a pod. They were extremely fit. The pathologist made some interesting observations. All lived in a warm climate. They had the allover suntans. They had heavily calloused feet. Hand callous patterns were the same without any occupational indicators. Their haircuts were probably from the same barber and the dental work was similar.

Still no federal records were found on them. Of course, there was the possibility they were from another country. But the rental agents maintained that the two men that they saw were pure Americans by looks, speech and mannerisms.

The investigation died a rather rapid, silent death for an event that took seven and maybe eight lives.

Epilogue

My name is Bolek Baca. At the time of the attached investigation, I was a Lieutenant with the Sheriff's Office. I became involved in this affair early on and then followed the various events through until it was declared a "Closed Case" in the "Unsolved File."

The demise of this case was not because of lack of viable leads or areas of potential investigation, but because of political pressure. Senator Lansing Liston became involved in the case due to the murder of his son, Larami, by my father, Ladoslav Baca. The details of that crime are fully covered in the sheriff's reports.

It quickly became obvious that Liston was the great avenging angel. He swooped down out of Washington in a Navy jet and took command of the investigation. It was not by direct pointing of a finger and saying, "you do this or you do that." It was much more sinister. It was through money and coercion that the clean-up was accomplished.

Files disappeared, photos vanished, witnesses lost their memory and personnel were reassigned to other cases. These actions were not designed to cover up the tragic death of Larami Lipton. On the contrary, Senator Lansing Lipton was intent on creating a history that surely should lead to sainthood after the requisite number of years.

No, the Florida suppression was to eliminate any possibility of the Senator's name ever being connected with the case involving the death of the founders of the "A Few Good Old Men" movement and the Bird List."Since the murderer of Larami Lipton committed suicide, there was no individual upon whom the Senator could vent his wrath. The next best thing was to destroy the progenitors of the philosophy that had propelled the murderer against his son in an attempt to influence the Senator.

I am uncertain as to who avenged whom. Logic dictates the supposition that the six home invaders at Chucky Bentley's apartment were federal agents of some nature under the direction of Senator Lipton since he was the only interested federal entity involved in the case.

We figure the six men were home invaders since they were armed with illegal weapons in Chucky's apartment late at night. There were indications from the computer reconstruction that information was being put on the extra external hard drives that were present.

It is not a difficult assumption to make that the six military types did not go to Chucky's to kill themselves. From the tools found in the kitchen and near the outside propane tank, which did not blow, it would appear the team was there with ill intent.

The best scenario that fits the facts is that the six-some were there to retrieve computer data and then kill Chucky and Merv in an "unfortunate, accidental fire."

However, it is my contention that Chucky and Merv found their Final Mission and took the bad guys with them. I believe Chucky had filled one of his fiberglass stands with the c4 from the wrappers I found in the backyard. That form of explosive is widely used in the military so Chucky would be familiar with it. Although we never found the source of the c4, it shouldn't have been all that difficult for Chucky to find.

Although no pieces were recovered from the ruins, I would guess that the magneto and switch from the flashlight were used to set off a blasting cap necessary to ignite the c4

The concentration of force and the distribution of damage places Chucky at the center of the blast. This would be consistent with the supposition that, indeed, Chucky initiated the event.

It should be noted that the five invaders who were killed were never identified. Their bodies were not claimed. They were disposed of by the county. There is no further information available concerning the man who disappeared from the hospital. This is just one of many dead ends in this case.

However, the die was already cast. A number of different drums took up Chucky and Merv's beat in different parts of the country. There were no parades or marches because a major part of the plan was to operate on an individual basis so as not to be traceable or involve anyone else in the mission. The only item to bind adherents together was a two-letter label found somewhere on the body or in the vicinity. It was "FM" for Final Mission.

It seemed as an auxiliary wing evolved "A Few Good Old Man" spun off "A Few Good Old Women."

An elderly grandmother, with a walker, crossed the street from her house to blow away her grandson's drug dealer before turning a .45 hog leg on herself. At the morgue, they found "FM" scrawled across her stomach in lipstick.

The administration didn't want to acknowledge the movement, so little fanfare was made of any of the missions. The media handled the events as isolated news items.

However, the Internet picked up the slack and carried the word. Several blog sites became message boards for the culture. Locals reported on what the national media ignored. There were small homegrown retaliations that wouldn't have qualified for network coverage, but the blogs enumerated them.

Another factor was added into the mix. The baby boomers now swelled the candidate pool.

It generally became hazardous to push drugs. Sell to the wrong person and some complete stranger might come by and blow you away. Anyone with gray hair is looked upon with suspicion by those who carry guilt.

It wasn't long before the politicos started watching their rearview mirrors with concern. Both elected officials and appointees suffered losses because of actions they had taken. As time went along, political ideas became the causes of action. Politicians became less inclined to support unpopular measures. The "Bird List" became a commonly used term. The Mad Hatters were having a great time spreading their first Amendment rights.

The hot button issues such as abortion and stem cell research resulted in casualties on both sides. The numbers were not great but the effect was substantial.

The off-year election was remarkable in that many long serving members of the House and Senate decided to do something else. Conservatives made surprising gains in both houses. However, conservatives were not immune to scrutiny and they suffered losses too.

There was a gathering storm as the regular elections neared. The Progressives saw their best chance in decades of fundamentally changing the structure of the country fading away. Their leadership pressed ahead despite heavy public disfavor.

Add to this, no one was paying much attention to the financial health of the country. The big spenders and the wasteful ones began to feel the weight of A Few Good Old Men/Women.

It is time that I bundle this report up and stick it in a safe deposit box. Deputy Hal Martin is the only one who knows about this package. He keeps trying to get me to publish the case but Dad is still too close. The day may come when I will be able to deal with this material, but if not, someone will realize just how important this little bit of history really is.

At the last general election, the Progressives finally gained enough power to state publicly their intentions to fundamentally change the United States. The general population had developed a growing distrust of Washington's ability to reflect the will of the people. There were a host of changes needed to bring several aspects of the government under control.

The Democrats campaigned on change and were wildly successful. However, when they began implementing their programs of change, the people began to realize the Progressive concept of "fundamental change" was Socialism/Communism.

By the time this shift in direction became evident, the Progressives were deeply entrenched with diligent work on all fronts. The left could forge ahead with complete disregard of the majority view. This was done through legislation, executive orders, court activism, new regulations and all the sundry routes that were open.

The opposition had insufficient numbers to change anything. Their only game was harassment until another element came upon the scene. That new element started in a small American Legion Post when a half a dozen legionaries decided to right a few old wrongs demonstrating that there are still consequences for bad behavior. From this start, "A Few Good Old Men" grew.

It was my father who expanded the concept to include consequences for politicians acting contrary to the will of the vast majority of the people.

Another weapon was added to the public arsenal by these same legionnaires. It was called the "Bird List." It was a database covering those politicians who voted contrary to the consensus public interest or violated the public trust in a grievous manner. Chucky and Merv started the list to help the public find miscreants and those they hold dear, so that the unhappy citizens could express their feelings with actions of their choosing. Much of the fun of being a powerful Senator was removed by getting the "bird" or other expression of derision every time he appeared in public. The Senator's family and friends appreciated the same attention even less.

Chuck and Merv must have either had a warning or premonition that their time was short because they developed the list and then immediately farmed it out to the Mad Hatters, who took it and ran with it.

Armed with these new weapons, the public mounted an active response to their perceived grievances. To ignore the will of the people carried its perils. Also injustices can carry their own brands of pain.

It will take some time for the new reality to sink in. But I believe that it will.

Somewhere down the line, someone will want to know what sparked the rescue of the United States.

My final addition to this file will be a couple of headlines following the next general election. I hope one will be, "We, the People Win" and the other will be, "Progressivism is Dead." I believe the chances are pretty good.

ISBN   978-0-9820044-3-2